T0400012

foundations of **primary mathematics education**

foundations of **primary mathematics education**

an introduction to curriculum, pedagogy and content

2nd edition

John West & Fiona Budgen

Routledge
Taylor & Francis Group
LONDON AND NEW YORK

First published 2019 by Allen & Unwin

Published 2020 by Routledge
2 Park Square, Milton Park, Abingdon, Oxon OX14 4RN
605 Third Avenue, New York, NY 10017

First issued in hardback 2021

Routledge is an imprint of the Taylor & Francis Group, an informa business

 A catalogue record for this
book is available from the
National Library of Australia

Internal design by Romina Panetta
Set in 12/17 pt Adobe Garamond Pro by Midland Typesetters, Australia

ISBN 13: 978-0-367-71813-8 (hbk)
ISBN 13: 978-1-76052-969-7 (pbk)

Contents

Chapter 1

Introduction

Welcome to the second edition of *Foundations of Primary Mathematics Education*. This book has evolved from a need to provide pre-service teachers with subject knowledge in mathematics alongside current knowledge and beliefs about teaching and learning. This is reflected in the choice of title, *Foundations of Primary Mathematics Education*, rather than simply *Foundations of Primary Mathematics*. Many pre-service teachers admit to feeling unsure about the mathematics they will have to teach in primary school. Others comment on the difficulties they have in recognising how the theories of teaching and learning they meet in other parts of their course can be applied when teaching mathematics. Part A of this book briefly outlines many of the current beliefs about effective teaching and learning. You will investigate these in more detail in other parts of your teacher education course, so the purpose here is to show how they relate to the teaching and learning of mathematics.

Part B provides the mathematical content knowledge that you will require as a primary teacher. The content of this section extends beyond the primary level to about Year 9 of the Australian Curriculum. While you may not have to teach this content, knowing it is a key part of being a strong teacher of primary mathematics. An understanding of where the primary content is leading will help you to teach with more accuracy and purpose. In order to avoid confusion, content that goes beyond the primary domain has been identified throughout Part B with this icon:

In order to graduate, all pre-service teachers must pass a literacy and numeracy test known as LANTITE (Australian Council for Educational Research [ACER], 2017). As this is a test for all teachers, it is not exclusively focused on primary-level mathematics. The extended range of mathematical content in this book will assist pre-service teachers with aspects of their subject knowledge that may require attention.

The current context of mathematics education

According to the Australian Curriculum, Assessment and Reporting Authority (ACARA), 'learning mathematics creates opportunities for and enriches the lives of all Australians' (ACARA, 2017). In contrast, many students' experiences of school mathematics are profoundly negative. All too often the outcome of school mathematics is for students to develop an aversion to the subject that precludes them from achieving their full potential.

In 2013, the WA Department of Commerce commissioned a study of *Science, Technology, Engineering and Mathematics* (i.e., STEM) education in

WA schools. The study involved a review of existing research and interviews with representatives of the three education sectors as well as a broad range of organisations that provide support for STEM education in Western Australia. The study revealed that the achievement of Australian primary and secondary students in STEM subjects has declined significantly in the last two decades.

The *Trends in International Mathematics and Science Study* (TIMSS) is a comparison of student achievement in mathematics and science in different countries and is conducted every four years. The data obtained from international benchmarking studies such as TIMSS allow educational trends to be examined over time. For example, the mathematical achievement of Australian students has fallen from being significantly above that of the United States and England in 1995 to significantly below these countries in 2007 (Brown, 2009).

In 2007, Australian students were ranked 14th behind Hong Kong, Singapore, Chinese Taipei and Japan, followed by a group of eight European countries and the United States. Students in all of these countries achieved significantly higher average scores than those in Australia. Nevertheless, the mathematical achievement of Year 4 students in Australia remained above the international average and was significantly higher than that of 20 countries, including Sweden and New Zealand (Thomson, Wernert, Underwood & Nicholas, 2008).

Questions to consider

1. Why might Australian students' achievement in mathematics be declining?
2. What is it about the education systems in other countries that may account for the difference in student achievement?
   (If possible ask someone who has studied overseas.)
3. Why is STEM education an issue of such concern to:
   a. students b. teachers c. parents and d. employers?

By 2011, the Year 4 TIMSS achievement data revealed that Australia's ranking had fallen again, this time to 18th place, with Australian students being significantly outperformed by students from 17 other countries. Significant concerns were also raised about the very low proportion of Australian students reaching the advanced achievement benchmarks. According to TIMSS 2011, only 10% of Australian students reached the advanced benchmark compared to 43% of students from Singapore (Mullis, Martin, Foy & Arora, 2012).

Widespread concern about Australia's declining performance in STEM subjects has prompted calls from the Australian Chief Scientist for urgent action at national level (Office of the Chief Scientist, 2012). Despite this, Australia's achievement has continued to decline against our international competitors. Data from the 2015 TIMSS benchmarking study revealed that Australian students were ranked 28th in the world for mathematical achievement, having been *significantly* outperformed by 21 countries including Kazakhstan (International Association for the Evaluation of Educational Achievement [IEA], 2016).

In addition to the decline in achievement, TIMSS data has revealed that there is a significant decline in Australian students' attitudes towards mathematics between Year 4 and Year 8. There are also major disparities in student achievement levels, attitudes and rates of participation in STEM subjects according to socio-economic status, location, race and gender. Of particular concern in states such as Western Australia is the widening achievement gap between metropolitan and non-metropolitan students and between Indigenous and non-Indigenous students (Hackling, Murcia, West & Anderson, 2014).

Questions to consider

1. Why might student attitudes towards mathematics decline between Year 4 and Year 8?
2. What unique challenges does Australia face in ensuring that all students have access to quality STEM education?

Participation rates in advanced or specialised mathematics subjects declined in all Australian states and territories between 1991 and 2007. For example, in Western Australia the participation rate in Calculus was 13.9% in 1992, which dropped to 7.7% by 2007 (Ainley, Kos & Nicholas, 2008). It has been suggested that part of the reason for this decline was that advanced mathematics was no longer a prerequisite subject for a range of university courses.

The Programme for International Student Assessment (PISA) provides data on the comparative performance of 15-year-old students in the OECD countries. PISA studies are conducted every three years. The 2012 PISA data revealed that the achievement gap between the highest 25% and lowest 25% of Australian students, based on their socio-economic status, is equivalent to 2.5 years of schooling. A similar achievement gap exists between Indigenous and non-Indigenous students, while a lesser gap exists between metropolitan and regional students (Thomson, De Bortoli & Buckley, 2013).

There are also concerns that the proportion of Australian university graduates in STEM fields is lower than in leading Asian economies. Over the ten-year period from 2001 to 2010, the proportion of Australian undergraduate students enrolled in STEM courses fell from 23.7% to 18.8%. In comparison, 64% of students in Japan, 52% in China, 40.6% in South Korea and 33% in Russia are enrolled in STEM courses (Office of the Chief Scientist, 2012).

In a survey of Australian secondary schools, Harris and Jensz (2006) found that three out of four schools reported difficulty in recruiting suitably qualified mathematics teachers. McConney and Price (2009) found that there is a much higher incidence of out-of-field teaching in poor communities, rural and remote schools and 'hard to staff' metropolitan schools, and that this is a major contributor to the relative underachievement of students in these schools.

Structure of this book

Many of the topics introduced in this book will be revisited and developed over the course of your primary education degree.

This chapter has reviewed the current context of STEM education in Australia. The next chapter explores the nature of teachers' work, notions of effective teaching and the professional standards that all teachers are required to meet. Chapter 3 addresses the complex topic of motivation, including the challenges faced by educators in overcoming mathematics anxiety and fixed mindsets. Behavioural, cognitive and social learning explanations of human behaviour are discussed.

A great deal of teacher time is spent on planning, a topic which is addressed in detail in Chapter 4. Chapter 5 'The Learning Environment and Building Relationships' and Chapter 6 'The Learning Process' deal with classroom management and student learning. Chapter 7 explores the structure of the Australian Curriculum: Mathematics, including the rationale and the content and proficiency strands. Strategies for 'Assessment and Reporting' are discussed in Chapter 8, including strategies used by teachers to gather diagnostic, formative and summative assessment data and for communicating this to parents. Chapter 9 explores the ways in which teachers make use of technology both in and outside of the classroom.

Part B of the book deals with specific aspects of mathematical content. Chapter 10 explores whole numbers, place value and operations. Chapter 11 discusses important fraction concepts and calculations and Chapter 12 reviews decimals and percentages. Chapter 13 addresses measurement, while Chapter 14 reviews geometry. Chapter 15 provides an introduction to number patterns and sequences, which form the basis for developing algebra skills. Chapters 16 and 17 discuss statistics and probability, respectively.

Language plays a crucial role in the teaching and learning of mathematics. This book contains a glossary of the mathematical terms used in the primary years. The way in which terms are used in mathematical contexts often differs from their conventional usage. It is vital that teachers model correct usage of mathematical terminology and this begins with a strong foundation of content knowledge.

Part A
mathematics teaching

Chapter 2
Understanding the Profession

Teachers play a significant role in shaping the society of the future. There is no doubt that teachers are highly influential role models for students. For this reason, teachers are expected to meet the highest professional and ethical standards.

Teachers' work

The best teachers have a well-developed personal understanding of the structure and principles of the mathematics that they are teaching (Haylock & Manning, 2014). They display confidence in their mathematical content knowledge and their ability to teach mathematics to children.

A primary school teacher's work requires the planning, teaching and assessment of mathematics. This involves setting appropriate learning goals, selecting activities and resources that will support learning, answering children's questions, and identifying and addressing their errors or misconceptions. In order to do this, teachers must have a sound understanding of the mathematics that the children are learning.

The most effective teachers will relate their subject knowledge to their pedagogical knowledge; that is their understanding of how children learn at different ages and how to support that learning. This synthesis of content knowledge and pedagogy is known as pedagogical content knowledge (PCK). For a teacher of primary mathematics this means knowing what conceptions and preconceptions children are likely to bring with them, knowing which concepts are likely to be easy or difficult for students to learn and knowing how to build a concept using resources, explanations and demonstrations so that it is comprehensible to children of different ages and backgrounds. The key message here is that effective pedagogical content knowledge is dependent upon a deep understanding of primary-level mathematics concepts and skills.

Activity

Many people think that being a primary teacher is easy. There is a well-known adage, 'Those who can, do. Those who can't, teach.' Shulman (1986) re-wrote that adage as, 'Those who can, do. Those who understand, teach' (p. 14).

1. What knowledge would you consider is essential to a deep understanding of primary-level mathematics concepts and skills?
2. Why is knowledge of mathematics learning in the pre- and post-primary years essential to the development of a deep understanding of primary-level mathematics concepts and skills?

Overcoming mathematics anxiety

It is extremely common for students to develop a dysfunctional relationship with mathematics as a result of their schooling. For many people, the study of mathematics leads to anxiety, a sense of powerlessness, and erosion of their sense of self-efficacy. According to Westwood (2000), 'many intelligent people after an average of 1500 hours of instruction . . . still regard mathematics as a meaningless activity for which they have no aptitude . . . to think that all our effort has led to a situation of fear and loathing is depressing' (p. 31).

Ideally, students should come to view mathematics as a tool that they can use with confidence and understanding as a result of their schooling. The way in which teachers relate to mathematics is extremely important since students learn from observation and imitation. According to Bandura (1989), students are more likely to learn behaviours that are perceived as beneficial.

Teachers who themselves have had negative experiences of mathematics can unconsciously model anxiety, fear or avoidance behaviour to their students. The effect can be to perpetuate a cycle that can profoundly limit students' future learning opportunities and, ultimately, their career choices. The teacher has a crucial role to play in modelling a positive approach to mathematics.

Questions to consider

1. How do you think your personal experiences in mathematics might influence the way you approach the teaching of mathematics?
2. Think about your personal feelings during school maths lessons. Did you feel anxious or secure in the classroom? In hindsight, what did your teacher do (or what could they have done) to create the conditions necessary for learning mathematics with confidence?

There are a number of ways in which teachers can model a positive approach to mathematics. For example, teaching mathematics in authentic contexts can highlight the relevance and usefulness of the skills that are being taught. In addition, problem-solving activities allow students to draw together ideas from across the curriculum to create richer, more interesting problems to explore. In this way, students can begin to develop an appreciation of mathematics as a valuable problem-solving tool.

In recent years there has been increasing awareness that mathematics anxiety is an all too common outcome of school mathematics. Mathematics anxiety has been recorded in students as young as five years of age and it can develop into a debilitating, lifelong condition (Boaler, 2016). When students feel stressed, working memory becomes blocked and students cannot access the mathematical facts they know. Thus the onset of timed testing accounts for the beginning of maths anxiety in approximately one-third of students (Boaler, 2014).

Neurological research has revealed that mathematics anxiety activates both fear and pain responses in the brain. When students' brains are busy regulating their emotions, their performance and efficiency, even on simple tasks, can be impaired. Indeed, highly mathematics-anxious individuals perceive a 'subjective feeling of visceral threat' when confronted with a mathematical task (Artemenko, Daroczy & Nuerk, 2015, p. 2). Increased activation in the pain perception network is observed when a student is presented with a mathematical task but not during the task itself, explaining why mathematics-anxious individuals try to avoid mathematics. While doing mathematical tasks, highly mathematics-anxious individuals experience increased activation in the amygdala, which is the part of the brain known for fear perception. Its activation during a mathematical task confirms that these students are afraid of mathematics (Artemenko et al., 2015).

As noted in Chapter 1, there are concerns about Australian children's declining levels of achievement in mathematics. Some of this decline has been attributed to teaching standards. It is true that many primary teachers feel

Questions to consider

1. How does understanding that mathematics anxiety is associated with pain and fear responses affect the way you choose to approach teaching mathematics?
2. People are sometimes able to overcome their fears or phobias by a process of progressive desensitisation. How might this work in mathematics?

insecure in their knowledge of the mathematics they have to teach. This has resulted in some schools attempting to assist teachers by purchasing commercial resources that can be implemented across all year levels. Rather than supporting teachers to become better teachers of mathematics, the result is often that teachers rely on these schemes to make their decisions about teaching for them and fail to provide 'careful, systematic and appropriate explanation of mathematical concepts, procedures and principles' (Haylock, 2006, p. 1).

Effective teaching

There is a great deal of evidence to suggest that one of the most important factors affecting student learning is the effectiveness of the classroom teacher (Hattie, 2012). Effective teachers have a rich knowledge of the content that they need to teach and appropriate strategies with which to teach it. They can create and maintain a safe and supportive learning environment. An effective teacher can improve the learning outcomes of all students in their class; this is particularly the case in mathematics (Hattie, 2016).

Effective teachers constantly strive to enhance their practice by using research to inform their instructional choices (Sullivan, 2011). In recent years there have been many developments in the educational and neurological research that have implications for the teaching and learning of mathematics in the classroom. By seeking out appropriate professional learning opportunities, teachers can ensure that their students benefit from the latest research on how students learn mathematics.

Some teachers also choose to participate in research studies as a way of improving their effectiveness. Partnerships between schools and universities have several benefits. Working with universities ensures that schools benefit from the latest research. University researchers who work in association with schools can focus on important issues that affect teachers and students. Collaboration between schools and universities allows pre-service teachers to benefit from examples of best practice that occur in schools.

Questions to consider

1. What does it mean to be an effective/ineffective teacher of primary mathematics?
2. What are the characteristics of effective/ineffective teachers?
3. How can a teacher use research to improve their effectiveness?

In the current educational climate, teachers' work and what happens in schools is under a great deal of scrutiny from various stakeholders. The Australian Institute for Teaching and School Leadership (AITSL) devised a set of seven standards for Australian teachers in an effort to articulate the expectations of stakeholders in the education profession.

Australian professional standards for teachers

The Australian Professional Standards for Teachers (AITSL, 2014) provide an outline of what teachers should be able to do at each of four career stages: graduate, proficient, highly accomplished and lead teachers. Teachers need to be aware of their professional obligations under the Australian Professional Standards.

Reading

Australian Institute for Teaching and School Leadership. (AITSL, 2014). *Australian Professional Standards for Teachers.*
www.teacherstandards.aitsl.edu.au/static/docs/Australian_Professional_Standard_for_Teachers_FINAL.pdf

The seven standards are divided into three domains: *Professional Knowledge* (Standards 1 and 2), *Professional Practice* (Standards 3, 4 and 5) and *Professional Engagement* (Standards 6 and 7). Within each standard there are several focus areas.

STANDARD 1: Know students and how they learn

STANDARD 2: Know the content and how to teach it

STANDARD 3: Plan for and implement effective teaching and learning

STANDARD 4: Create and maintain supportive and safe learning environments

STANDARD 5: Assess, provide feedback and report on student learning

STANDARD 6: Engage in professional learning

STANDARD 7: Engage professionally with colleagues, parents/carers and the community

Professional knowledge

Standard 1 requires teachers to have a fundamental knowledge of child development and the psychology of learning. Chapter 3 of this text discusses various attempts to explain the motivation of human behaviour as it relates to learning. Chapter 6 discusses theories of the way in which knowledge is constructed.

Standard 2 requires teachers 'know the content and how to teach it'. For a primary school teacher, this requires knowledge of mathematical content up to and including Year 7 level and an understanding of how this content is taught to students (i.e., mathematical pedagogy). The Australian Curriculum: Mathematics (ACARA, 2017) describes the rationale, sequence and structure of the mathematics content taught in Australian schools. Teachers must understand the structure and purpose of the Australian Curriculum in order to teach effectively. The Australian Curriculum is discussed in detail in Chapter 7.

Professional practice

Standards 3, 4 and 5, respectively, address the planning, teaching and assessment of learning. The role of the teacher requires short, medium and long-term planning. Teachers spend a great deal of time planning individual lessons (and lesson components), the development of a topic over a series of lessons (perhaps over several weeks) and developing teaching programs (in which topics are developed over an extended period such as a term or year). Planning is addressed in more detail in Chapter 4.

Teachers also need to know how to create and maintain a safe and supportive learning environment within their classroom. Students learn most effectively when they feel secure in their learning environment. The creation of a safe and supportive learning environment is discussed in Chapter 5.

Teachers are also responsible for assessing and providing feedback on student learning. As schools are legally obliged to provide parents with feedback on student progress, formal assessment procedures are used to collect and report data on student progress. However, teachers collect assessment data in a variety of ways and for a variety of reasons. These include providing students with individualised feedback on their progress, for making planning decisions and also to determine the effectiveness of their own teaching. Assessment and reporting strategies are addressed in Chapter 8 of this book.

Professional engagement

Teachers are also expected to plan for their ongoing skill development. Standard 6 requires teaches to engage in relevant professional learning opportunities. Continued professional learning enables teachers to hone their existing skills and to develop new ones.

The many stakeholders in the educational community include students, parents, teachers, administrators, librarians, cleaners, gardeners, employers, local businesses, school psychologists and medical practitioners. The teacher's role requires them to be able to work collaboratively as a member of a team. There are several organisations at local and state levels that provide support for teachers; the national parent organisation is identified below. Many offer discounted membership to pre-service teachers and first-year out teachers.

Professional associations

Every state and territory has a professional association for primary and secondary mathematics teachers. These associations organise student activities and competitions as well as professional development for teachers. They usually have an extensive library of mathematical resources and hold conferences for primary and secondary teachers.

The Australian Association of Mathematics Teachers (AAMT)
www.aamt.edu.au

Joining an affiliated state association provides automatic membership to the AAMT. The association provides access to a range of free online resources and organises various student activities and competitions. AAMT also provides professional development for teachers and has a range of books and teaching resources for sale (at discounted prices for members).

Teacher registration

In order to register as a teacher in Australia, candidates must have completed all of the requirements of a nationally accredited teacher education program. Initial teacher education programs are accredited by the teacher regulatory authority in each state or territory.

In recent years, increasing public concern about the levels of literacy and numeracy among initial teacher education students led the Australian Government to introduce mandatory literacy and numeracy testing. The Literacy and Numeracy Test for Initial Teacher Education students (LANTITE) is a single instrument designed to demonstrate that graduates of all Australian teacher education programs have personal levels of literacy and numeracy within the top 30% of the population (ACER, 2017). All students who commenced a teacher education course after 1 January 2017 are required to pass the test prior to graduation (TRBWA, 2016).

The LANTITE focuses on the mathematics that teachers at all levels may be required to use in their daily and professional lives. As such, it extends beyond the mathematics that is taught in the primary curriculum. The following table is provided as an indicative (but not exhaustive) list of the mathematical content likely to be included in the LANTITE test.

For more information about LANTITE visit:
- https://teacheredtest.acer.edu.au/

Numeracy area	Example content
Number and Algebra	Proportional reasoning; ratio; fractions (including score conversions); percentages (including weighted percentages across assignments); decimals; scientific notation; money; budgeting; interest calculations; basic operations; simple formulae; calculation of GST
Measurement and Geometry	Time; timetabling and scheduling (e.g., parent–teacher interviews, timetables across multiple campuses); knowledge about space and shape, symmetry and similarity relevant to common 2D and 3D shapes; quantities, including areas and volumes; use of given relevant routine formulae; conversion of metric units; use of maps and plans, scales, bearings
Statistics and Probability	Interpreting mathematical information such as graphs; statistics and data (including NAPLAN data); comparing data sets or statistics; statistics and sampling, including bias; distributions; data and interpretation validity; reliability; box plots—matching data to displays; actual against predicted scores; assigning a grade based on a raw score; interpreting/calculating an ATAR; drawing conclusions about student achievement based on data.

Australian Council for Educational Research (ACER), 2017

This chapter has explored the nature of teachers' work, particularly as it relates to the teaching and learning of mathematics. It has looked at the issue of mathematics anxiety and how building confidence is a crucial aspect of becoming an effective teacher. It has also discussed the national professional teacher standards and teacher registration.

Chapter 3
Motivation

Motivation is the process that activates, guides and maintains behaviour (Krause, Bochner & Duchesne, 2003). Psychologists have sought to explain human behaviour for many decades. Various explanations of behaviour have been proposed and these are outlined in the sections below. Teachers need to understand what motivates their students as this has a significant impact on planning and classroom management.

Questions to consider
1. What do you think motivates children to learn mathematics?
2. How can teachers motivate children to learn mathematics?
3. What role does personal motivation play for the teacher?

Behavioural explanations

At the most basic level, behavioural explanations for learning concern the connections or associations that link a stimulus and a response (i.e., a reaction to a stimulus). As early as the 1890s, Edward L. Thorndike studied learning in animals by placing cats in puzzle boxes and seeing how long it took them to escape to receive a food reward. From this he proposed that any behaviour that is followed by pleasant consequences is likely to be repeated, whereas behaviour that is followed by unpleasant consequences is likely to be stopped. He called this idea the Law of Effect.

In the 1930s, B.F. Skinner built on the work of Thorndike. Skinner believed that the best way to understand animal behaviour was to look at the causes of an action and its consequences. According to Skinner, positive reinforcement strengthens a behaviour by providing a consequence that is rewarding. A behaviour can also be strengthened by removing an unpleasant consequence; this is known as negative reinforcement.

Punishment is the opposite of reinforcement since it weakens or eliminates behaviour. It can involve applying an unpleasant consequence or removing a pleasant consequence. The use of punishment is generally best avoided since it does not necessarily guide towards the desired behaviour, can cause increased aggression and undesirable behaviours often return when punishment is no longer present.

Behaviourists argue that learning means producing a particular response to a particular stimulus and that behaviour can be shaped by reinforcement. There are several aspects of mathematics that benefit from this approach to learning. For example, learning basic addition facts and multiplication tables requires learning specific responses to particular stimuli. Another example of a learned behaviour is when students respond to a question by raising their hand.

Cognitive explanations

Cognitive theorists argue that motivation stems both from the desire to accomplish goals and the need to avoid failure. Achievement motivation was defined by Atkinson and McClelland as a stable personality characteristic that drives some individuals to strive for success (Krause et al., 2003). Students with high levels of achievement motivation are motivated to attempt a task if they believe they are likely to be success-

ful. In contrast, students with low levels of achievement motivation are more likely to attempt tasks which are very easy and have little risk of failure, or tasks which are very hard so that failure will not be considered their fault. This theory also explains why some students do not always try to be successful. Students will avoid tasks when their need to avoid failure is greater than their need for success.

Attribution theory explores the way in which an individual's explanation of success and failure influences their subsequent motivation (Krause et al., 2003). Students with high achievement motivation tend to attribute their success or failure to internal factors such as ability or hard work (in the case of success) and inadequate preparation or lack of effort (in the case of failure). Students with low achievement motivation, who have a need to avoid failure,

Questions to consider
1. How might the way in which students attribute their success and failure contribute to a state of learned helplessness? (i.e., when students feel helpless to avoid negative situations)
2. How might the way in which students attribute their success and failure lead to hostility towards the teacher?

tend to attribute their success to external factors beyond their control such as good luck or good teaching (in the case of success) and bad lack or poor teaching (in the case of failure).

Atkinson's Expectancy-Value Theory states that the effort that a student is willing to invest in a task is the product of two factors. These are:

- the degree to which the student expects to be successful in the task if they make an effort, and
- the perceived value of doing the task or the outcomes to be gained from completing the task successfully.

Both factors must be present if a student is to undertake a task. If students believe that they are unable to succeed, they are unlikely to invest effort, regardless of the extent to which the student values completing the task. Similarly, if the student does not value the task, effort will not be invested regardless of how likely they might be to succeed (Barry & King, 1998).

Activity

Consider the implications of expectancy-value theory for the design of mathematical learning experiences. According to the theory, what characteristics should classroom activities possess in order for students to invest effort in them?

Social learning theory explanations

Albert Bandura is a Canadian psychologist whose ideas incorporate elements of both behaviourist and cognitive approaches. He is widely regarded as one of the most influential psychologists of all time. Bandura's social learning theory recognised the contribution that mental factors play in the learning process. Bandura (1989) argued that behaviour is the result of a complex interaction between cognitive, behavioural and environmental influences, and that people learn from each other by observation, imitation and modelling.

A significant limitation of the behaviourist approach is that explanations arising from animal studies neglect the influence of human cognition. Behavioural theorists such as Thorndike were therefore unable to explain the more complex

aspects of human behaviour (such as reading and problem solving). According to social learning theory, people not only learn by being rewarded or punished, they can also learn from watching somebody else being rewarded or punished (i.e., observational learning). The theory suggests that human thought plays an important role in deciding whether an individual will imitate behaviour. For example, students will be more likely to imitate a behaviour if the perceived rewards outweigh the perceived costs. When students observe teachers modelling appropriate behaviour it can encourage them to behave in the same way. Teachers can therefore use role modelling to shape children's behaviour in ways that are beneficial to learning and development. For example, teachers can model a collaborative approach to problem solving.

Questions to consider

1. What are the implications of social learning theory for teachers in the mathematics classroom?
2. What inappropriate behaviours relating to mathematics may students observe and imitate in the classroom?
3. What kinds of mathematical skills can be learned and taught through observation, imitation and modelling?

Humanist explanations

Abraham Maslow was an American psychologist who developed one of the most widely recognised theories of motivation. Maslow developed a five-tier model based on consideration of human needs. Maslow's theory assumed:

- that human needs are never completely satisfied
- behaviour is motivated by the need for satisfaction, and
- needs can be classified in a hierarchical structure.

Maslow argued that once a particular need was satisfied, it no longer served as the primary motivator for human behaviour. Maslow's hierarchy is often presented as a pyramid, with the most basic needs being physiological (e.g., oxygen, food and water). Beyond this the individual is motivated to seek basic security, stability, protection and freedom from fear. Once these needs have been satisfied, the need for belonging and love emerges, followed by the need to develop self-esteem, confidence and status. Assuming all of the previous needs have been met, an individual feels the need for self-actualisation. That is, to be actively engaged in achieving their full potential.

For example, a student is unlikely to develop confidence in their mathematical problem-solving ability if they do not feel safe in the classroom environment. Similarly, students whose physiological needs have not been met (for example, if for whatever reason they have not eaten breakfast) will not be motivated to perform well in lessons.

Mathematical mindset

Mindset is an idea arising from the work of Professor Carol Dweck of Stanford University. Dweck has studied achievement motivation in an extensive research career that has spanned several decades. Dweck discovered that many adults believe that certain basic qualities (such as intelligence and mathematical ability) are simply fixed traits. They also believe that it is talent alone that creates success. This is known as a fixed mindset. In the context of mathematics education, children with a fixed mindset believe that nothing can change their abilities in mathematics and, when they meet a problem they find difficult, they are likely to attribute the difficulty to inadequate ability. Dweck found those who believe that their basic abilities can be developed through dedication and hard work have what she identified as a growth mindset (Dweck, 2017). In the context of mathematics education, children with a growth mindset believe that they can become better and they are more likely to practise skills and persevere with challenging tasks.

Professor Jo Boaler (also of Stanford University) argues that there is an urgent need to overhaul the teaching and learning of mathematics in schools. While mathematical problems arise in almost all aspects of everyday life, few students leave school with an appreciation of mathematics as a powerful problem-solving tool (Boaler, 2009). Instead many students leave school convinced that they do not have a so-called 'maths brain' and consequently that it will be impossible for them to ever experience success in mathematics (Boaler, 2015).

Boaler has built on the work of Dweck by applying an understanding of mindset to the teaching and learning of mathematics. A rapidly growing body of research literature has attested that students who adopt a growth mindset are more likely to succeed in mathematics at all levels (Boaler, 2016). Teachers who adopt a growth mindset in their teaching are careful not to pre-judge students' potential for achievement. They believe that all students can further develop their abilities with effort. This is not to say that some students do not have more mathematical ability than others; rather that *all* students can improve their skills once they dispense with limiting notions such as the 'maths brain'.

> ### Questions to consider
>
> There are no quick fixes or short cuts to developing a deep understanding of primary-level mathematics concepts and skills.
> 1. In what ways will the teacher's mathematical mindset affect the development of their mathematical understanding?
> 2. What impact might relying on commercial teaching resources have on the development of a teacher's mathematical understanding and pedagogical content knowledge?

Classroom implications

Motivating students to engage in learning is an essential aspect of the teacher's role. Poor motivation often leads to low achievement and can result in classroom management problems. Psychological research on motivation has provided teachers with insight into the internal processes that serve to  activate, guide and maintain student behaviour. The behavioural, cognitive, social learning and humanist approaches have each contributed to our understanding of motivation in the classroom.

Activity

Complete the table below by identifying some examples of classroom behaviour that are best explained using the behavioural, cognitive, social learning and humanist approaches. For the purpose of this activity, focus on the types of behaviour that might be seen in the context of mathematics lessons.

Approach	Behaviour(s) apparent in mathematics lessons
Behavioural	<ul><li></li><li></li><li></li><li></li></ul>
Cognitive	<ul><li></li><li></li><li></li><li></li></ul>
Social Learning	<ul><li></li><li></li><li></li><li></li></ul>
Humanist	<ul><li></li><li></li><li></li><li></li></ul>

Which of the approaches would be most appropriate for
a. *activating*, b. *guiding* and c. *maintaining* student behaviour?

Chapter 4

Planning

Good teaching rarely happens by accident. Planning is an integral aspect of teachers' work. Careful planning ensures that a teacher has a clear idea of what, when, how and why students are going to learn. Planning also allows teachers to anticipate problems before they occur in a lesson. Planning can help build teacher confidence and reduce anxiety.

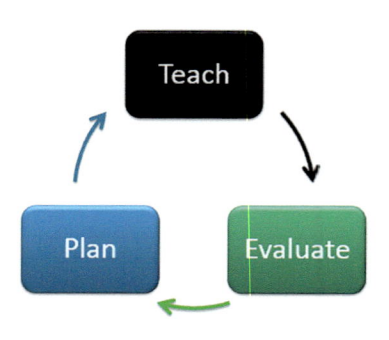

Teachers are responsible for the learning program that occurs in their classroom and as such they are expected to carefully document their planning. Planning occurs at several distinct levels, depending on the timeframe that is being considered. Teachers plan for the short, medium and long term, and often planning happens at several levels simultaneously.

The backward planning process

For teachers, the starting point in the planning process is usually to identify what it is that students need to have achieved at the end of the teaching and learning cycle. The teacher can then identify an appropriate assessment strategy to assess whether the learning outcomes have been achieved. This is known as *backward planning*.

In backward planning, the desired learning outcomes are used as the starting point to select appropriate student learning experiences. Typically, a teacher will begin with a yearly plan, which is then broken down into term plans, which are divided into smaller units of work. During the planning phase, teachers need to make decisions not only about what content to teach and how to sequence it, but also time allocation, teaching resources and classroom organisation (Booker, Bond, Sparrow & Swan, 2004).

Yearly planning

Yearly planning (sometimes known as *programming*) is usually conducted at the beginning and end of the school year. The purpose of the yearly program is to provide a broad overview of the work to be completed during each term. The yearly program should address all aspects of the relevant

achievement standard (see Chapter 7 for more information) and include a balance of all curriculum strands i.e., number and algebra, measurement and geometry, and statistics and probability.

Term planning

Once teachers have established an overall plan for the year, this is broken down into plans for each individual term. The term plan provides more detail about the topics to be covered and the sequence in which they will be taught. The term plan is usually

broken down into units of work that are generally from two to four weeks in duration.

Weekly and daily planning

Short-term planning occurs at the level of individual lessons or learning activities. Short-term planning usually happens in the days immediately before the learning experience occurs and provides the finer details of the learning outcomes, learning activities, timings, materials and resources to be used. Teachers often work from a daily work pad and their planning should consider how they will organise students and groups in the classroom and how they will assess what has been learned.

Lesson planning

There are many things that teachers need to consider when planning lessons. The first step is generally to identify the learning intentions. In other words, what is it that you want students to achieve as a result of the learning experience?

The learning intentions are generally written in lesson plans as *Specific Learning Goals*. These are developed from the *Content Descriptions* in the mathematics curriculum. The Content

Descriptions outline the knowledge, skills and understandings in mathematics that teachers are expected to teach at every year level. All Content Descriptions are supported by one or more *elaboration* to illustrate and clarify the learning that should take place, as shown in the example below:

Mathematics Strand and Year Level	Number and Algebra, Year 5
Sub-strand	Fractions and decimals
Content Description	Recognise that the place value system can be extended beyond hundredths
Elaborations	• using knowledge of place value and division by 10 to extend the number system to thousandths and beyond • recognising the equivalence of one thousandths and 0.001

This is one of four Content Descriptions relating to fractions and decimals in the Year 5 mathematics curriculum. Student mastery of each Content Description is developed over a series of lessons, with each lesson having one or two Specific Learning Goals related to the Content Description. A Specific Learning Goal of one of these lessons might be to 'Identify and record mixed numbers with denominators of 10, 100 and 1 000 as decimals.'

Consider how the Specific Learning Goal, 'Identify and record mixed numbers with denominators of 10, 100 and 1 000 as decimals', has been written. Specific Learning Goals need to be expressed in terms of what the student will be able to demonstrate on completion of the lesson. In the example above, the verbs *identify* and *record* make it clear what the students need to do in order to show their understanding of the concept. Once the teacher has defined the skill, knowledge or attitude that he or she wants the students to demonstrate, the step-by-step process of planning how to help the students to get there and how to assess their success can begin.

When planning the lesson steps, the teacher should consider the knowledge, concepts or skills that need to be taught and the types of learning experiences that he/she will provide. The lesson introduction should link the new material to the students' prior knowledge and experiences. The subject matter needs to be broken down into small parts and the body of the lesson should contain carefully sequenced learning activities that promote the desired learning. The lesson conclusion should provide students with a sense of closure. It should draw together the ideas, skills and concepts that were the focus of the lesson and allow students to reflect on their learning.

The teacher must also consider the teaching resources or materials that can be used to support students in their learning. In mathematics lessons, this may include calculators, rulers, protractors, spinners, dice, playing cards, blocks and grid paper. Lessons may also involve the use of computers, tablets or smart phones. In these situations, the teacher may need to consider how equipment will be distributed and packed away in an efficient manner.

Lesson planning involves making decisions about assessment. In order to evaluate the success of a learning experience, teachers must select an assessment strategy that will reveal what, and how much, individual students have learned. A balanced program will include a range of assessment strategies that enable all students to show their learning. Lesson planning also involves making reasonable learning and teaching adjustments for students with special educational needs. This may involve working with an educational assistant (EA).

Finally, lesson planning should consider the structure of the learning environment. Will students work individually or in groups? Will students be working on the mat, in the kitchen area, outside or at their desks? Chapter 6 of this book discusses various ways in which teachers can manage the learning environment.

Building number sense

Sparrow and Swan (2005) suggest beginning each mathematics lesson with a brief mental session. The purpose of this is both to develop students' number sense and to establish a routine. Simply calling out a series of unrelated questions may 'alienate children and create a classroom climate of

fear' (Sparrow & Swan, 2005, p. 41). You might choose to focus on a particular calculation strategy or ask a series of related questions that emphasise a key understanding.

Beginning a lesson

There are many ways teachers capture students' interest and 'tune them in' at the beginning of a lesson. Some examples include:

- Solve a puzzle
- Show a picture
- Use trivia
- Read a children's book
- Watch a <u>brief</u> video clip
- Use a real-life scenario
- Solve a riddle
- Tell a story
- Play a short, focused game
- Use some Mathemagic
- Ask a 'thinking' question
- Listen to a song
- Do a brainstorm
- Ask a hypothetical question.

Activity

Choose four of the lesson starter ideas above. Develop each one into a ready-to-use 'tuning in' activity. For example, if you choose to read a children's book, identify some suitable books for teaching a particular mathematical topic.

The body of the lesson

You will usually have between 20 and 40 minutes of on-task time for students to come to grips with the mathematics being taught. Have a clear purpose in mind so you don't become side-tracked. The body of the lesson will generally involve individual, pair or group work. Clear instructions at the outset will help avoid time wasted repeating instructions. Set clear parameters for the lesson such as how equipment is to be used, how long the task should take and the quality of work expected. During this phase of the lesson, monitor how the children's learning progresses. Observe how they handle the task or tasks that you have set to see that they stay on task. You may be required to clarify instructions, give further explanations or ask students questions. You may note that some children find the task too simple or too difficult, in which case you should be prepared to offer a modified version of the task.

Some teachers run several groups within their class. Their knowledge of the students allows them to group students of similar ability for certain lessons. At other times, different groups may be used that allow for peer-tutoring among the children. Rather than plan a different lesson for each group, choose a common topic and modify to meet the needs of various groups within the class. At times you may plan several different activities to form the basis of work stations. Groups will then rotate through these work stations over a series of lessons. Sometimes you will opt for this approach for pragmatic reasons such as not having enough equipment for everyone to do the same activity at the same time.

Do your best to visit most, if not all, of the children during the body of the lesson. Look for specific things (e.g., the questions that students ask you and each other) that will allow you to judge how well individuals and the class as a whole are coping with the work. This will form part of your evaluation of how well the lesson went.

Concluding a lesson

Often the conclusion of a lesson is rushed because the body of the lesson runs overtime. The conclusion is a vital part of the lesson and should be thoroughly planned. Generally, the whole class should be brought together at the end of a lesson and key parts of the lesson reviewed. Use this time to gauge how well the lesson went. What did the children learn as a result of participating in the lesson? Individuals or groups might report on their findings or you might review the key points of the lesson. It is during this time that planning for any follow up lessons begins, as you will have a better 'feel' for the children's knowledge and understanding of the topic.

Planning your explanation

Many beginning teachers underestimate how difficult it can be to give clear explanations of mathematical ideas to primary school-age students.

Activity

Observe your mentor teacher explaining a specific mathematical concept to students. Think about the choice of words he or she uses. What aspects of their explanation did they emphasise? Were any aspects unclear to students? Did the students have questions? If so, how did the teacher respond?

The key to a good explanation is to plan it thoroughly. Effective teachers spend a great deal of time planning their explanations using language that students will understand. When planning your explanation, you should take note of any aspects of the topic that are likely to cause difficulty. In order to do this, teachers consult research on the teaching and learning of the particular topic. You should give illustrations or examples to make difficult points clear and these examples should be included on your lesson plan.

You should also ensure that your explanation uses age-appropriate language. While mathematical vocabulary includes many familiar English words, these words often have distinct meanings when used in a mathematical context. When giving your explanation, you should monitor students for signs of understanding or confusion. Even the best explanation in the world can be derailed if you have not first gained students' full attention.

Activity

There are many common English words that have particular meanings when used in a mathematical context. Make a list of common English words that have different meanings in mathematics compared to their conventional meaning.

Planning focus questions

Teachers will often use focus questions as a technique for gaining instantaneous feedback on student understandings and/or misunderstandings. In addition to planning their explanations, teachers will often prepare a handful of questions that focus on the most important key aspects of the topic(s) being discussed. Asking questions is one way of engaging with students, keeping their attention and reinforcing their participation.

Questions can be used to stimulate discussion and activate students' critical and creative thinking. The use of carefully selected focus questions also allows the teacher to emphasise important aspects of a topic. By anticipating the types of questions that students might ask in the planning stage, the teacher can plan their explanation accordingly. Teachers should ensure that the focus questions are productive, usually open-ended and do not serve to side-track the class.

Planning for diversity

Planning for diversity focuses on how learning can be adapted, or *differentiated*, to accommodate students with a diverse range of learning needs. The Australian Association of Mathematics Teachers (AAMT) includes in the range those who are:

- English as an added language or dialect students
- high achievers and gifted students
- students who have special needs, and
- students at risk in keeping up with their learning.

While teachers often find differentiation one of the more challenging aspects of their role, there are many benefits to diversity in the classroom. Students and teachers in diverse classrooms are exposed to a broader range of issues and perspectives and gain an increased capacity for tolerance, respect and concern for others. Students and teachers have opportunities to re-evaluate their personal perspectives and values and gain a clearer idea of how information is processed in different cultures. Students learn in an environment which is more representative of the community in which they live and learn new ways of collaborating in class.

Differentiated teaching of mathematics provides for the wide range of students found in every classroom. It has already been noted that, rather than planning different lessons for each group, a common topic can be chosen and modified to meet the needs of various groups within the class. This is differentiating the content of the lesson. Teachers can also differentiate the processes in the lesson by varying how students go about the set tasks, and the products of the lesson by considering opportunities for students to present what they have learned in different ways. The success of differentiated mathematics learning depends on teachers who believe that all students have the capacity to succeed. It is true that some students will show more aptitude for mathematics than others, but all students can achieve competence.

Recent developments in assistive technologies (such as tablets and iPads) provide significant benefits for many students with developmental, physical, sensory or cognitive impairments. With appropriate guidance from teachers, the use of these technologies allows students to function more autonomously in the classroom. This can reduce the load on educational assistants (EAs) and allow students to participate more fully in learning activities.

Planning as a professional obligation

It can sometimes appear to beginning teachers that their mentors do little in the way of formal planning. Nevertheless, planning is a professional obligation and the expected mental preparation for teaching. Registered teachers have demonstrated their planning ability and are able to provide the necessary documentation on request.

The simplest and most effective test of the adequacy of teacher planning is whether another educator could teach from your plan without the need for further elaboration. In situations where you are unable to perform your duties (such as illness or jury duty), it is a typical expectation that you would provide lesson plans or guidance as to what your students have learned and what they need to learn next. For this reason, it is recommended that all teachers have a number of 'backup' lessons on standby that can used at short notice.

Reflecting on planning

An important aspect of lesson planning is to reflect on the phases of the learning experience that worked well and those that require refinement. In hindsight it is usually easy to identify the effective aspects of a lesson and those that require modification. Reflective teachers use their self-evaluation of previous lessons as the starting point for planning subsequent learning experiences. For students, this can help to provide a sense of continuity between lessons. For teachers, it enables them to anticipate any issues and incorporate them in their planning of subsequent learning experiences. Flexibility is an essential element of all lesson planning since it is impossible to plan for all possible contingencies.

Some things to consider when reflecting on your mathematics lessons are:

- Was I well enough prepared with all the materials and resources I needed?
- Were my explanations adequate and clear? Who needed additional explanation and why? Who has mastered the ideas and who will need more time? How will I assist the students who need more help?
- Were my instructions clear? Did students know what they had to do?
- Were the transitions between activities smooth?
- Did I make the best use of the time available?

The following page contains a typical learning experience plan template that a pre-service teacher on practicum would be expected to complete and show to their mentor teacher prior to teaching each lesson.

Learning Experience Plan			
Learning Area: Strands:	Date:	Time:	Year Level:
Content Description:			
Specific Learning Goals:			
Achievement Standard:			
Assessment: **What** will you monitor?		Recording: **How** will you monitor?	
Students' Prior Knowledge:			

Time:	Teaching and Learning Strategies: Introduction:	Focus Questions:	Preparation/ Resources:
	Body:		
	Conclusion:		

Learning Experience Plan			
	Learning and Teaching Adjustments:		

To be completed AFTER the learning experience

Evaluation of Students' Learning:

Self-Evaluation:

Chapter 5

The Learning Environment and Building Relationships

An important outcome of schooling is that students learn to function effectively in a social environment. Students must learn to work cooperatively with others and resolve conflicts when they occur. In this respect, teachers are important role models. The learning environment that the teacher creates can shape the knowledge, behaviour, skills and values of students. In particular, the learning environment in which students learn mathematics affects their conception of what mathematics is, how it is used and how it is learned (Bobis, Mulligan & Lowrie, 2013). If teachers are to counter negative attitudes towards mathematics, prevent anxiety about mathematics and enable children to attempt challenges without fear of embarrassment, then they must create a safe and supportive learning environment in the classroom.

Both students and teachers have important roles to play in creating a classroom environment conducive to mathematics learning. By setting appropriate boundaries for behaviour, the teacher contributes to students' sense of security in the classroom environment. At the beginning of the school day, the teacher assumes  a duty of care to maintain the safety of their students. The students understand that it is their job to learn while it is the teacher's job to ensure a safe environment in which learning can occur.

An effective classroom management style enables a teacher to build positive relationships and create a healthy learning environment. There are many different approaches to managing children's behaviour and it is important that pre-service teachers adopt an approach that is consistent with their school's ethos and policy. For the purposes of consistency, pre-service teachers should generally adopt a behaviour management approach that is similar to that of their mentor teacher(s). This chapter outlines some of the different approaches to managing and guiding children's behaviour in order to build a classroom environment in which mathematics learning can flourish.

Kounin's instructional management

Jacob Kounin's approach to classroom management was to focus on instructional management. He analysed the effect that certain teacher behaviours have on the behaviour of students in classrooms. Kounin's research indicated that both student behaviour and learning are influ-  enced by the way in which teachers organise and present lessons, arrange the classroom, attend to students and make transitions between lessons (Charles, 2011).

Kounin's research identified a number of characteristics (or traits) displayed by effective teachers. These included:

- Knowing what is happening in all parts of the classroom at all times (Kounin called this teacher trait 'with-it-ness').
- Seamless lesson presentation (a trait Kounin called 'smoothness').
- Having systems for gaining student attention and clarifying instructions.
- Keeping students attentive and actively involved.
- Ensuring that learning activities are both enjoyable and challenging.

The instructional management techniques described by Kounin are helpful in creating and maintaining a classroom environment that is conducive to learning. By keeping students smoothly engaged in overlapping learning activities, teachers can greatly reduce instances of inappropriate behaviour in the classroom. Effective teachers also use low-key responses to prevent misbehaviour from escalating. For example, the teacher could move closer to a student who is talking out of turn (i.e., proximity) or shake their head to indicate that a particular behaviour is not appropriate (i.e., visual cue). Kounin's approach underscores the value of effective lesson planning (Charles, 2011).

Activity

How would you apply Kounin's characteristics of effective teachers when teaching mathematics? Consider what it means for lesson planning, organising resources and knowing what your students can (and cannot yet) do.

Ginott's congruent communication

Haim Ginott was born in Israel in 1922 and was a classroom teacher in the early part of this career. He worked as a professor of psychology and a consultant psychologist for the *Today* show in the United States. Haim Ginott's research focused on the way in which communication between teachers and students affects behaviour (Charles, 2011). Ginott suggested that when dealing with problem behaviour, it is important to focus on the behaviour and not the character of the student. Teachers should speak to students in a manner they would expect to be spoken to themselves and invite cooperation by focusing on the desired behaviour, rather than the misbehaviour.

Ginott felt that teachers have the power to make or break a student's self-confidence and that this power is wielded primarily through communication. He argued that communication between teachers and students should address *situations* and not the student's character and personality. In this way, class-room discipline is gradually attained through small, gentle steps rather than strong tactics (Charles, 2011).

Self-confidence is important in mathematics learning and Ginott's approach can help to protect this by putting the focus on the student's actions and not their personality or characteristics. Undesirable behaviour is sometimes exhibited when students wish to avoid attempting a task for fear of embarrassment or failure. Some students prefer to create a distraction and face the consequences rather than reveal that they are struggling with the content of a lesson. In these situations, the teacher needs to consider the motivation behind the student's actions.

Dreikurs' democratic teaching

Rudolf Dreikurs was an Austrian psychologist and educator who felt that schools are places where young people are nurtured and helped to feel that they belong and are valued. Dreikurs argued that all students want to belong, and that misbehaviour was the result of feeling a lack of belonging to one's social group. When students feel a lack of belonging in the classroom, they behave in ways designed to gain attention, gain power, exact revenge or gain sympathy (Charles, 2011).

It is not unusual for students to feel a lack of belonging in mathematics lessons. Students who are having difficulty with the mathematics content being taught may feel isolated from the rest of the class. They may seek to gain attention and favour from their peers through humour, defiance or distraction. For this reason, teachers need to ensure that students feel that their presence is valued and that they do not feel ignored or powerless. For example, the teacher may group students of similar ability together for tasks to foster a sense of belonging.

Glasser's noncoercive discipline

William Glasser was an American psychiatrist and educational consultant. He believed that quality should be the prime focus of teaching, learning and curriculum. According to Glasser, teachers should provide a warm and supportive classroom environment, ask students to only do work that is useful and always ask students to do the best that they can (Charles, 2011).

Glasser believed that all humans have a need for survival, belonging, power, freedom and fun and that most misbehaviour is the result of boredom and frustration. He advocated the use of classroom meetings as a regular part of the curriculum. He also suggested that teachers learn non-punitive and non-coercive techniques for motivating students to work and participate in the classroom (Charles, 1999).

Questions to consider
1. Should all mathematics teaching be restricted to what is useful?
2. How important is it for students to have the freedom to have fun with numbers such as exploring patterns, playing games and learning calculation tricks?

Kohn's beyond discipline

Alfie Kohn argues that traditional approaches to classroom management and discipline focus on eliciting student compliance rather than helping students become caring and responsible human beings. According to Kohn, the most effective teachers rely on collaborative problem-solving instead of coercive control. Kohn suggests that teachers who wish to move beyond traditional models of discipline must provide an engaging program, develop a sense of community and involve students in decision-making.

Kohn believes that classrooms should operate more like communities. He describes communities as places in which students feel cared about and are encouraged to care about others. Kohn advocates the use of classroom meetings and encourages teachers to involve students in all aspects of problem-solving (including managing behaviour). According to Kohn, students develop self-control when teachers show trust (Charles, 2011).

Questions to consider

1. What do you think are the essential elements of an engaging mathematics curriculum?
2. How can encouraging students to care about others assist in creating a classroom environment in which mathematics learning can flourish?

Developing a personal approach

The quality of the learning environment is dictated by the extent to which students and teachers feel safe and secure in the classroom. Building effective relationships in the classroom is arguably the most important aspect of teaching. This chapter has provided a brief overview of five different approaches to building relationships in the classroom. Each of these approaches contributes to our understanding of the complex interaction between teacher and student behaviour.

From a risk management point of view, mathematics lessons are generally safe *physical* environments; for example, there is no risk of spilled chemicals and little risk of cuts or bruises. Nevertheless, in our work with pre-service teachers we have encountered many adults who can attest that, when  handled poorly, children's mathematical experiences can result in significant *emotional* trauma and lasting negative impacts on confidence and self-esteem. This should be an important consideration when deciding on a classroom management approach.

Children are more likely to display negative attitudes and disruptive behaviour when they are confused, frustrated or insecure in the learning environment. Students who feel that they are valued members of the class who are expected to be held accountable for their own learning are more likely to behave in a responsible manner. Students who are actively engaged in enjoyable and productive learning will often exert peer pressure on class members who disrupt their activities (Booker et al., 2004).

There is a great deal of useful advice for teachers who wish to develop an effective classroom *persona*. When it comes to teaching mathematics, three of the five personal qualities listed by Barry and King (1998) stand out:

1. *Be pleasant.* Students respond accordingly when teachers are relaxed, friendly and enthusiastic. Be positive and constructive. Teachers cannot build lasting relationships that are harsh, negative or destructive.
2. *Be approachable.* It is important students feel they can approach the teacher, not as a threatening adult, but as someone who has the time to hear them out.
3. *Be tolerant.* Teachers who are not bothered by the differences between students are more likely to be viewed as fair and consistent. Students have a strong sense of social justice and notice when teachers stray

from being fair and reasonable. The reality is that teachers will not like everyone in the classroom to the same extent, but it is essential that teachers behave in a manner that is accepting of everyone.

Activity

Complete the table below by identifying elements of the listed behaviour management approaches that you might consider incorporating into your personal behaviour management style when teaching mathematics.

Approach	Ideas to incorporate
Jacob Kounin's Instructional Management	
Haim Ginott's Congruent Communication	
Rudolf Dreikurs' Democratic Teaching	
William Glasser's Noncoercive Discipline	
Alfie Kohn's Beyond Discipline	

Chapter 6

The Learning Process

Mathematics education is not only about learning to read, write and do mathematics. Teachers are also highly influential in the social and emotional development of students during their formative years. It is important therefore that teachers understand the academic, social and emotional growth of their students.

This chapter begins by considering the nature of teaching and learning in mathematics and then reviews frameworks that have been proposed to explain how knowledge is constructed. Finally, it explores some different strategies for managing the physical and social environment in the classroom and how resources can be used to support student learning.

What is learning?

Teaching and *learning* are often used synonymously when in fact there is a world of difference between the two terms. For example, it is almost certain that at some point in your education you were *taught* some mathematics that you didn't *learn*. That isn't necessarily a reflection on your mathematical ability—maybe your teacher didn't explain it well, or you were distracted by the student sitting next to you at the time.

Learning is generally considered to be a positive thing, although you can probably think of situations where someone has learned a bad habit. When we *learn*, we acquire new or modify our existing knowledge, behaviours, skills, values or preferences. Learning may lead to changes in the way we approach or react to new or familiar situations. There is still a great deal about the learning process that we don't completely understand.

Piaget's stages of cognitive development

Jean Piaget was a Swiss psychologist known for his pioneering research into child development. Piaget argued that the learner plays an active role in adapting to their environment. This adaptation begins at birth and proceeds through a series of four distinct stages. These are the *sensorimotor stage* (approximately 0–2 years), the *preoperational stage* (approximately 2–7 years), the *concrete operational stage* (approximately 7–11 years) and the *formal operations stage* (approximately 11 years plus). The characteristics of each stage are outlined below.

Students in a primary classroom will not necessarily all be operating at the same level, so it is helpful to ascertain the cognitive levels of the students and adjust teaching accordingly. Piaget believed that the amount of time children spend at each stage could vary, depending on their environment and their personal characteristics (Ojose, 2008). While it is not possible to teach cognitive development specifically, teachers can provide an environment that fosters and encourages higher levels of thinking.

Stage	Ages	Description	
Sensorimotor	0–2 years	Infants construct knowledge and understanding by associating sensory information and physical interactions with objects. A characteristic that children at this stage develop is their ability to link numbers to objects e.g., one teddy, two cars, three blocks. This is the foundation of counting.	
Preoperational	2–7 years	Children begin to learn to speak but cannot manipulate information mentally. The child has difficulty seeing events from the viewpoint of others. During this stage, children develop formal counting skills and the ability to answer questions such as 'Who has more?' or 'Are there enough?' Children at this stage may have trouble reversing operations. For example, a child at this stage who understands that adding three more to five gives eight may not be able to reason that taking three from eight leaves five. Children at this stage generally only consider one attribute at a time so, for example, if liquid is poured from a long, narrow container where the level is high into a wider container where the level is lower, they believe that there is less liquid in the second container.	

Concrete Operational	7–11 years	Thought processes become more mature. Children are able to draw inferences from observations to make generalisations. Children are still developing deductive reasoning skills. Children at this stage are able to consider several attributes simultaneously so in the liquid example above, the child would notice that the container is wider and understand that if nothing has been taken away or added then the quantity of liquid must be the same, even though the level is lower. 'Hands-on' activities are vital at this stage to explore abstract mathematical concepts such as place value and arithmetical operations, in order to make them meaningful.
Formal Operations	11 years +	At this stage, children become capable of hypothetical and deductive reasoning. They develop the ability to think about abstract concepts (e.g., variables) such as those used in science and mathematics. The formal operational learner is typically able to reason abstractly so can understand and perform arithmetical operations purely by using symbols. The learner can solve problems such as $x + 2x = 12$ without the problem being contextualised in a word problem such as, 'Aldo served a certain number of customers. Rob served twice as many. How many customers did Aldo serve?'

Vygotsky's sociocultural theory

Lev Vygotsky (1896–1934) was a Russian psychologist who argued that cognitive development is essentially a social process. Vygotsky believed that it is through social interactions with others that children learn how to think. According to Vygotsky, knowledge must first be constructed in a social and cultural context before it can be adopted by individuals. A child's thinking is therefore shaped by their interactions with others in a way that is culturally appropriate (Krause et al., 2003).

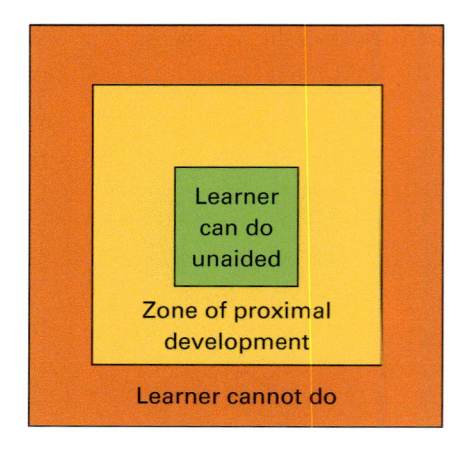

Vygotsky identified the *zone of proximal development*, which he defined as the distance between what a child can do unassisted and what they can do with the support of others (Krause et al., 2003). The assistance provided by others is known as *scaffolding*. Scaffolding occurs in many forms; it may involve providing structure to a task or directing attention to the relevant aspects. In some cases, it may involve providing hints. In others, it may involve asking appropriate questions. Scaffolding may come from peers as well as the teacher.

Constructivist learning theory

Piaget also pioneered the idea that children actively construct their own learning as a result of their experiences. This idea, which gained prominence in the 1960s, remains a major influence on educational approaches used today, particularly in subjects that have a highly hierarchical nature (such as mathematics and science) (Krause et al., 2003). According to Piaget, individuals *assimilate* new information when they incorporate it into an already existing framework. Individuals must reframe their mental representation of the world in order to *accommodate* new experiences. Accommodation can therefore be seen as the process by which failure leads to learning. The results of a growing number of neurological studies reinforce the notion that making mistakes is an important component of learning in mathematics (Boaler, 2016).

The constructivist approach that is in common use today has also been strongly influenced by the work of Vygotsky. The *social constructivist* approach views cognition as a collaborative social process involving interactions with the environment and self-reflection. According to social constructivists, students are not passive recipients of information transmitted by the teacher; rather they are active participants in the learning process. In this model, the instructor assumes the role of a facilitator whose role is to help students to develop (or construct) their own understanding.

Constructivism encourages educators to recognise the value of the knowledge and experience that students bring to learning. As individuals interact with others and their environment, they link information learned through their experiences to previous knowledge and so construct new understandings and new knowledge. The teacher's role in this approach is to provide learner-centred experiences that enable students to build on their existing knowledge. They should also provide opportunities for students to work together and learn from one another. The teacher should also provide scaffolding to assist novice learners to develop expertise (Krause et al., 2003).

Bloom's taxonomy

Benjamin Bloom published his *Taxonomy of Educational Objectives* in 1956 and it is still widely used in education today. Bloom's taxonomy classifies educational learning objectives into various levels of complexity. A revised version of the taxonomy was published in 2001.

The cognitive domain is broken down into six levels. The lowest of these levels simply involves rote memorisation of existing knowledge. The second level (understanding) requires comprehension of the main ideas. The third level (applying) entails the use of acquired knowledge to solve problems in new situations.

Analysing involves breaking information down into its component parts and determining how these parts relate to one another. Evaluating includes the presentation and defence of opinions by making judgements about information, the validity of ideas or quality of work based on a set of criteria. The highest level (creating) involves bringing parts together to form a cohesive whole.

CREATE

EVALUATE

ANALYZE

APPLY

UNDERSTAND

REMEMBER

Teachers should ensure that learning activities provide opportunities for students to develop higher-order thinking. This can be done by providing a variety of learning experiences, using different instructional groupings and ensuring that learning activities do not merely focus on the memorisation of facts. There are various ways of arranging the learning environment to ensure that students have opportunities to engage with different levels of complexity.

Whole-class instruction

Whole-class instruction is a useful technique when teachers need to convey instructions or information to their entire class. Teachers will often use this approach when introducing a new concept, or at the beginning or end of a lesson. Before commencing instruction, the teacher should ensure that they have gained students' full attention. Teachers should also be mindful not to raise their voice to talk over students, as this can put a strain on their voice and can be taken as tacit acceptance of student behaviour.

While whole-class instruction can be an efficient means of communicating information, like broadcasting, it is primarily a one-way medium. Since the teacher is in charge, students are required to accept less responsibility for their own learning. As a result, whole-class instruction should be used sparingly.

Instructions should also be carefully planned so that the teacher can anticipate the kinds of questions that may occur or clarifications that may be needed (Reys et al., 2012).

Small group work

Most teachers use a great deal of small group work in their classrooms. Small group work enables students to learn from one another and generally provides more motivation than working on individual tasks. The use of small groups also allows teachers to cater for students with diverse learning needs. In planning small group work, the teacher needs to decide whether students will be placed in homogeneous (i.e., the same) or heterogeneous (i.e., different) ability groups.

Another benefit of small group work is that interaction between students assists in the development of their communication skills. Students must also learn the skills required to work cooperatively, such as turn-taking and sharing. The teacher can assist by modelling strategies for dealing with disagreements and different points of view.

In relation to mathematics instruction, small group work is often used in conjunction with problem-solving tasks. These tasks typically require students to apply a range of strategies or approaches since the solution to the problem is not immediately obvious. The teacher has an important role to play in monitoring small group work and providing guidance where necessary. It is important to avoid the temptation to simply provide students with the answer to a problem-solving task.

Individual work

There are times when the teacher or the nature of a particular task requires students to work individually. Some students enjoy setting their own goals and working at their own pace. Other students find it difficult to self-regulate their behaviour, but this is a skill that can be improved with practice. When students are working individually, the teacher is free to provide individual assistance. An alternative is to set up a contract system in which students agree to complete certain tasks over a set period of time, but can choose the order in which they complete the tasks (Booker et al., 2004). The use of contracts may be more appropriate with older (or more independent) students.

Questions to consider

Think about the social aspects of the learning environment that you intend to create in your classroom.

1. How will the learning environment encourage students to explain their ideas and justify their thinking?
2. How will the learning environment encourage students to make sense of the mathematics they do?
3. How will you encourage students to discuss mathematics from their everyday experiences?
4. What will you do to encourage students to work independently and collaboratively to solve problems?
5. How will you ensure that students are provided with sufficient time to explore new materials and concepts?

Adapted from Bobis, Mulligan and Lowrie (2013)

Using games

Games are an important element of most mathematics programs for many reasons. Firstly, games are intrinsically motivating and provide an enjoyable and efficient way of reinforcing mathematical skills. Games are also highly flexible; they can generally be modified to fit a particular timeslot and can be played as a whole-class, in small groups or individually. An example of a mathematics game is shown opposite.

For the teacher, incorporating a chance element into a game serves a dual purpose—the study of probability is an integral part of the mathematics curriculum and games are not always won by the strongest mathematicians

NOGGLE

Place nine 10-sided dice in a 3 $\times$ 3 partitioned box and shake the box to generate a set of random numbers. Use **all** of the numbers with a combination of addition, subtraction, multiplication and division to make a target as close as possible to 50.

7	3	2
1	4	9
2	4	6

There may be many ways in which a solution can be found. For example:

First add (4 + 4 + 2) = 10. Multiply by 2 (i.e., 20).
Add (7 + 3) $\times$ (9 − 6) = 30. This leaves us the digit 1 to use.
Multiply (20 + 30) $\times$ 1 = 50.

We could also try (7 + 3 + 2) $\times$ (1 + 2) = 36. We can then add 9 + 6 = 15, giving 51. We can then subtract 4 ÷ 4 = 1, giving a total of exactly 50.

in the class. Games may therefore engage a broader range of students. Most mathematical games can be readily modified to provide appropriate levels of challenge for a wide range of students.

Many games require only the kinds of resources that are commonly found in classrooms (such as dice and counters). Games also teach important social skills such as turn-taking, communication and sharing. From a purely logistical point of view, playing games in groups greatly reduces the amount of equipment that is required, which can be an important consideration when planning.

Using manipulatives

Swan and Marshall (2010) defined a mathematical manipulative as 'an object that can be handled by an individual in a sensory manner during which conscious and unconscious mathematical thinking will be fostered' (p. 14). In deference to Piaget's developmental stages, mathematical manipulatives are sometimes referred to as concrete materials. There are a broad range of materials that can be used to develop mathematical ways of thinking. These include:

- Pattern blocks
- Base 10 blocks
- Attribute blocks
- Polydron/Geoshapes
- Unifix cubes
- Multilink cubes
- Cuisenaire rods
- Pop sticks
- Dice
- Counters
- Trundle wheels
- Miras

Safety

Safety is the most important consideration in managing the physical aspects of the learning environment. The classroom arrangement should be free from clutter, obstructions or trip hazards (such as extension cords or bags on

Activity

Each of the mathematical manipulatives identified below is appropriate for teaching particular mathematical topics. Identify topics for which each type of manipulative is appropriate.

Manipulative	Name	Topics
	Base 10 blocks	
	Cuisenaire Rods	
	Pattern Blocks	
	Geoboards	
	Dice (regular and specialist)	
	Miras	
	Magnetic Construction Kits	
	Trundle Wheel	

the floor). Chairs and desks should be arranged to ensure that teachers and students are not twisting their heads or bodies for prolonged periods. Heavy equipment should be stored safely. Classroom lighting should be bright, preferably natural, but should avoid direct sunlight or reflections shining into children's faces.

Questions to consider

Think about the physical aspects of the learning environment that you intend to create in your classroom.

1. Does the physical environment include a range of mathematical manipulatives that are clearly labelled and readily accessible to students?
2. Do students have their work on display around the room?
3. Does the learning environment have:
   — Spaces where children can work in groups?
   — Spaces that allow for discussion between students?
   — Spaces where children can work quietly?
4. Do students have ready access to a range of stimulus materials e.g., mathematical posters, books, timetables and newspapers?
5. What are the potential safety issues that you need to consider in order to ensure that you create a safe learning environment?

Adapted from Bobis et al. (2013)

Teachers must create a safe and supportive classroom environment if it is to be conducive to the teaching and learning of mathematics. This chapter has discussed various theories of instructional and classroom management. Teachers usually draw on a combination of their understanding of the theoretical approaches and their classroom experience in order to develop their personal approach to building relationships and managing the learning environment.

Chapter 7

Unpacking the Curriculum

The Australian Curriculum, Assessment and Reporting Authority (ACARA) is the independent statutory authority who has overall responsibility for the Australian Curriculum and the administration and reporting of the National Assessment Program—Literacy and Numeracy (NAPLAN). Writing of the Australian Curriculum: Mathematics commenced in 2009 and involved expert writers, advisory groups and extensive consultation with a range of stakeholders. The Australian Curriculum: Mathematics describes the essential mathematical content knowledge and skills that students should achieve as a result of their formal schooling. The rationale, descriptions of stages of learning, descriptions of the content and proficiency descriptions of strands, and the Year 7 achievement standard included below have been reproduced verbatim from the Australian Curriculum: Mathematics (ACARA, 2017).

The Australian Curriculum: Mathematics

Rationale

Learning mathematics creates opportunities for and enriches the lives of all Australians. The Australian Curriculum: Mathematics provides students with essential mathematical skills and knowledge in Number and Algebra, Measurement and Geometry, and Statistics and Probability. It develops the numeracy capabilities that all students need in their personal, work and civic life and provides the fundamentals on which mathematical specialties and professional applications of mathematics are built.

Mathematics has its own value and beauty and the Australian Curriculum: Mathematics aims to instil in students an appreciation of the elegance and power of mathematical reasoning. Mathematical ideas have evolved across all cultures over thousands of years and are constantly developing. Digital technologies are facilitating this expansion of ideas and providing access to new tools for continuing mathematical exploration and invention. The curriculum focuses on developing increasingly sophisticated and refined mathematical understanding, fluency, logical reasoning, analytical thought and problem-solving skills. These capabilities enable students to respond to familiar and unfamiliar situations by employing mathematical strategies to make informed decisions and solve problems efficiently.

The Australian Curriculum: Mathematics ensures that the links between the various components of mathematics, as well as the relationship between mathematics and other disciplines, are made clear. Mathematics is composed of multiple but interrelated and interdependent concepts and systems which students apply beyond the mathematics classroom. In science, for example, understanding sources of error and their impact on the confidence of conclusions is vital, as is the use of mathematical models in other disciplines. In geography, interpretation of data underpins the study of human populations and their physical environments; in history, students need to be able to imagine timelines and timeframes to reconcile related events; and in English, deriving quantitative and spatial information is an important aspect of making meaning of texts.

The curriculum anticipates that schools will ensure all students benefit from access to the power of mathematical reasoning and learn to apply their mathematical understanding creatively and efficiently. The mathematics curriculum provides students with carefully paced, in-depth study of critical skills and concepts. It encourages teachers to help students become self-motivated, confident learners through inquiry and active participation in challenging and engaging experiences.

Activity

Highlight three points you consider to be most important in the rationale above. Share and discuss the reasons for your choices with others.

The aims of the Australian Curriculum: Mathematics are to ensure that students:

- Are confident, creative users and communicators of mathematics, able to investigate, represent and interpret situations in their personal and work lives and as active citizens.
- Develop an increasingly sophisticated understanding of mathematical concepts and fluency with processes, and are able to pose and solve problems and reason in Number and Algebra, Measurement and Geometry, and Statistics and Probability.
- Recognise connections between the areas of mathematics and other disciplines and appreciate mathematics as an accessible and enjoyable discipline to study.

> **Activity**
>
> Reflect on the above aims. Based on your personal experience, which of these aims do you feel were achieved in relation to your own learning in mathematics?

Mathematics across Foundation to Year 12

Although the curriculum is described year by year, the document provides advice across four stages on the nature of learners and the relevant curriculum:

- Foundation–Year 2: typically, students from 5 to 8 years of age
- Years 3–6: typically, students from 8 to 12 years of age
- Years 7–10: typically, students from 12 to 15 years of age, and
- Senior secondary years: typically, students from 15 to 18 years of age.

Foundation–Year 2

The early years (5 to 8 years of age) lay the foundation for learning mathematics. Students at this level can access powerful mathematical ideas relevant to their current lives and learn the language of mathematics, which is vital to future progression.

Children have the opportunity to access mathematical ideas by developing a sense of number, order, sequence and pattern; by understanding quantities and their representations; by learning about attributes of objects and collections, position, movement and direction, and by developing an awareness of the collection, presentation and variation of data and a capacity to make predictions about chance events.

Understanding and experiencing these concepts in the early years provides a foundation for algebraic, statistical and multiplicative thinking that will develop in subsequent years. These foundations also enable children to pose basic mathematical questions about their world, to identify simple strategies to investigate solutions and to strengthen their reasoning to solve personally meaningful problems.

Years 3–6

These years emphasise the importance of students studying coherent, meaningful and purposeful mathematics that is relevant to their lives. Students still require active experiences that allow them to construct key mathematical ideas, but also gradually move to using models, pictures and symbols to represent these ideas.

The curriculum develops key understandings by extending the number, measurement, geometric and statistical learning from the early years; by building foundations for future studies through an emphasis on patterns that lead to generalisations; by describing relationships from data collected and represented; by making predictions and by introducing topics that represent a key challenge in these years, such as fractions and decimals.

In these years of schooling, it is particularly important for students to develop a deep understanding of whole numbers to build reasoning in fractions and decimals and to develop a conceptual understanding of place value. These concepts allow students to develop proportional reasoning and flexibility with number through mental computation skills and to extend their number sense and statistical fluency.

Years 7–10

These years of school mark a shift in mathematics learning to more abstract ideas. Through key activities such as the exploration, recognition and application of patterns, the capacity for abstract thought can be developed and the ways of thinking associated with abstract ideas can be illustrated.

The foundations built in previous years prepare students for this change. Previously established mathematical ideas can be drawn upon in unfamiliar sequences and combinations to solve non-routine problems and to consequently develop more complex mathematical ideas. However, students of this age also need an understanding of the connections between mathematical concepts and their application in their world as a motivation to learn. This means using contexts directly related to topics of relevance and interest to this age group.

During these years, students need to be able to represent numbers in a variety of ways; to develop an understanding of the benefits of algebra, through building algebraic models and applications and the various applications of geometry; to estimate and select appropriate units of measure; to explore ways of working with data to allow a variety of representations and to make predictions about events based on their observations.

The intent of the curriculum is to encourage the development of important ideas in more depth and to promote the interconnectedness of mathematical concepts. Teachers will, in implementing the curriculum, extend the more mathematically able students by using appropriate challenges and extensions within available topics. A deeper understanding of mathematics in the curriculum enhances a student's potential to use this knowledge to solve non-routine problems, both at this level of study and at later stages.

Structure

The Australian Curriculum: Mathematics is organised around the intersection of three content strands and four proficiency strands. The content strands describe the mathematical content knowledge that is to be taught and learned. The proficiency strands describe how content is explored or developed and they have been incorporated into the content descriptions of the content strands. This approach has been adopted to ensure students' proficiency in mathematical skills develops throughout the curriculum and becomes increasingly sophisticated over the years of schooling.

	Understanding	Fluency	Problem Solving	Reasoning
Number and Algebra		■		■
Measurement and Geometry	■		■	
Statistics and Probability		■		

Content Strands

Number and algebra

Number and Algebra are developed together, as each enriches the study of the other. Students apply number sense and strategies for counting and representing numbers. They explore the magnitude and properties of numbers. They apply a range of strategies for computation and understand the connections between operations. They recognise patterns and understand the concepts of variable and function. They build on their understanding of the number system to describe relationships and formulate generalisations. They recognise equivalence and solve equations and inequalities. They apply their number and algebra skills to conduct investigations, solve problems and communicate their reasoning.

Measurement and geometry

Measurement and Geometry are presented together to emphasise their relationship to each other, enhancing their practical relevance. Students develop an increasingly sophisticated understanding of size, shape, relative position and movement of two-dimensional figures in the plane and three-dimensional objects in space. They investigate properties and apply their understanding of them to define, compare and construct figures and objects. They learn to develop geometric arguments. They make meaningful measurements of quantities, choosing appropriate metric units of measurement. They build an understanding of the connections between units and calculate derived measures such as area, speed and density.

Statistics and probability

Statistics and Probability initially develop in parallel and the curriculum then progressively builds the links between them. Students recognise and analyse data and draw inferences. They represent, summarise and interpret data and undertake purposeful investigations involving the collection and interpretation of data. They assess likelihood and assign probabilities using experimental and theoretical approaches. They develop an increasingly sophisticated ability to critically evaluate chance and data concepts and make reasoned judgements and decisions, as well as building skills to critically evaluate statistical information and develop intuitions about data.

Content descriptions

The mathematics curriculum includes content descriptions at each year level. These describe the knowledge, concepts, skills and processes that teachers are expected to teach and students are expected to learn. However, the content descriptions do not prescribe particular approaches to teaching. They are intended to ensure that learning is appropriately ordered and that unnecessary repetition is avoided. However, a concept or skill introduced at one year level may be revisited, strengthened and extended at later year levels as needed. Content descriptions are grouped into sub-strands to illustrate the clarity and sequence of development of concepts through and across the year levels. They support the ability to see the connections across strands and the sequential development of concepts from Foundation to Year 10. The table below shows the sub-strands in each of the three content strands.

Number and Algebra	Measurement and Geometry	Statistics and Probability
Number and place value (F–8)	Using units of measurement (F–10)	Chance (1–10)
Fractions and decimals (1–6)	Shape (F–7)	Data representation and interpretation (F–10)
Real numbers (7–10)	Geometric reasoning (3–10)	
Money and financial mathematics (1–10)	Location and transformation (F–7)	
Patterns and algebra (F–10)		

Content elaborations are provided for Foundation to Year 10 to illustrate and exemplify content and assist teachers to develop a common understanding of the content descriptions.

Proficiency Strands

The proficiency strands describe the various contexts in which mathematical content knowledge is taught. While not all proficiency strands apply to every content description, they indicate the breadth of mathematical actions that teachers can emphasise.

Understanding

Students build a robust knowledge of adaptable and transferable mathematical concepts. They make connections between related concepts and progressively apply the familiar to develop new ideas. They develop an understanding of the relationship between the 'why' and the 'how' of mathematics. Students build understanding when they connect related ideas, when they represent concepts in different ways, when they identify commonalities and differences between aspects of content, when they describe their thinking mathematically and when they interpret mathematical information.

Fluency

Students develop skills in choosing appropriate procedures, carrying out procedures flexibly, accurately, efficiently and appropriately, and recalling factual knowledge and concepts readily. Students are fluent when they calculate answers efficiently, when they recognise robust ways of answering questions, when they choose appropriate methods and approximations, when they recall definitions and regularly use facts, and when they can manipulate expressions and equations to find solutions.

Problem solving

Students develop the ability to make choices, interpret, formulate, model and investigate problem situations and communicate solutions effectively. Students formulate and solve problems when they use mathematics to represent unfamiliar or meaningful situations, when they design investigations and plan their approaches, when they apply their existing strategies to seek solutions and when they verify that their answers are reasonable.

Reasoning

Students develop an increasingly sophisticated capacity for logical thought and actions, such as analysing, proving, evaluating, explaining, inferring, justifying and generalising. Students are reasoning mathematically when they explain their thinking, when they deduce and justify strategies used and conclusions reached, when they adapt the known to the unknown, when they transfer learning from one context to another, when they prove that something is true or false and when they compare and contrast related ideas and explain their choices.

Achievement standards

The achievement standards summarise the mathematical content that students are required to have mastered by the end of each year level. The Year 7 achievement standard is shown below. By the end of Year 7, students will:

- solve problems involving the comparison, addition and subtraction of integers
- make the connections between whole numbers and index notation and the relationship between perfect squares and square roots
- solve problems involving percentages and all four operations with fractions and decimals
- compare the cost of items to make financial decisions
- represent numbers using variables
- connect the laws and properties for numbers to algebra
- interpret simple linear representations and model authentic information
- describe different views of three-dimensional objects
- represent transformations in the Cartesian plane
- solve simple numerical problems involving angles formed by a transversal crossing two parallel lines
- identify issues involving the collection of continuous data
- describe the relationship between the median and mean in data displays
- use fractions, decimals and percentages, and their equivalences
- express one quantity as a fraction or percentage of another
- solve simple linear equations and evaluate algebraic expressions after numerical substitution

- assign ordered pairs to given points on the Cartesian plane
- use formulas for the area and perimeter of rectangles and calculate volumes of rectangular prisms
- classify triangles and quadrilaterals
- name the types of angles formed by a transversal crossing parallel line
- determine the sample space for simple experiments with equally likely outcomes and assign probabilities to those outcomes
- calculate mean, mode, median and range for data sets, and
- construct stem-and-leaf plots and dot-plots.

Activity

Which aspects of the Year 7 achievement standard are you confident about explaining to another pre-service teacher? Highlight any terminology which you are unsure about and make a note of any mathematical content you feel you need to revise.

Glossary

The Australian Curriculum: Mathematics also includes an extensive glossary of mathematical terminology. Throughout the mathematics content descriptions, key mathematical terms are hyperlinked to their definitions in the glossary. A glossary of the terminology commonly used in primary school mathematics is included at the end of this book.

Scootle digital repository

Scootle is a growing digital repository of resources designed to support the teaching and learning of various mathematics concepts. The Australian Curriculum: Mathematics content descriptions are hyperlinked to resources in the

Scootle repository. Scootle can be accessed by registering as a teacher education student using your university email address. When you are browsing the resources, you need to look at the context and evaluate the resource's suitability for use with a particular group of students.

This chapter has reviewed the history, rationale and structure of the Australian Curriculum: Mathematics. It has looked at how mathematical content is structured and taught across the four stages of learning and the three content strands. It also discusses the four proficiency strands of understanding, fluency, problem solving and reasoning, which serve as the contexts within which the essential mathematical content is taught.

Chapter 8

Assessment and Reporting

There is a great deal of terminology associated with the measurement of student achievement. *Assessment* is the process of collecting and synthesising data about the knowledge, understanding and skills of a person or group. *Evaluation* refers to the judgements made in relation to the worth of a score, rating or ranking. *Reporting* is the process of documenting and communicating information about students' knowledge, understanding and skills to parents (Barry & King, 1998).

Principles of Assessment

There are a number of principles to which all assessment should adhere. The School Curriculum and Standards Authority of Western Australia provides a useful foundation to consider. It has defined six principles to reflect best practice in assessment. The principles are outlined below.

Assessment should be an integral part of teaching and learning

Teachers need to constantly monitor their students' progress and make adjustments to their teaching as required. Assessment should be based on multiple sources of evidence that, between them, provide sufficient measures of student learning. Assessment should incorporate a variety of strategies to gain a broad view of student learning. A typical approach to assessment could include marking assigned tasks, monitoring homework, diagnostic tasks and interviews, investigations, tests, quizzes and informal observation.

Assessment should be educative

A significant proportion of overall learning time is used for the purposes of assessment. While the primary goal of assessment is to provide feedback on student learning, assessment tasks should themselves contribute to furthering student learning. An assessment is educative if it is a context for learning; skilful teachers engage students in meaningful assessment tasks that in themselves are valuable learning experiences.

Assessment should be fair

An assessment task is fair if it provides all students with an equal opportunity to demonstrate their learning. Assessment tasks should be as free as possible from gender, racial, cultural or other potential bias. Appropriate accommodation must also be made for students with special needs.

Assessments should be designed to meet their specific purpose

This refers to the accuracy with which an assessment measures what it purports to. For example, the assessment of mathematical ability should not be confounded by the level of students' reading comprehension. That is not to say that there is no relationship between these skills; in fact, reading comprehension is extremely important in mathematics. Nevertheless, this is an important consideration in ensuring the assessment (and reporting) of mathematical ability is accurate.

Assessment should lead to informative reporting

It is a federal government requirement that schools report on student achievement using a range of grades from A to E. Grades are a broad classification of student performance and need to be supported by specific evidence that gives an accurate and objective account of the student's progress and achievement.

Assessment should lead to school-wide evaluation processes

Teachers and school leaders need to identify areas for improvement and be willing to evaluate both the intended and unintended consequences of any initiative to improve the target area.

For more information go to:
• https://k10outline.scsa.wa.edu.au

Types of Assessment

Assessment is divided into three broad categories based on the point in time when the assessment occurs and the purpose of the data being gathered. Diagnostic assessment generally occurs at the beginning of the teaching and learning cycle and provides the teacher with data about students' prior knowledge. Formative assessment involves gathering data during the teaching and learning cycle and allows the teacher to provide students with feedback that guides their subsequent learning. Summative assessment occurs at the end of the teaching and learning cycle. Its primary purpose is to gather data on which to make evaluative judgements about the extent of student learning. Each of these is discussed in more detail below.

Diagnostic assessment

In recent years, teachers have been held increasingly accountable for the learning outcomes achieved by their students. Diagnostic assessment can provide teachers with a reference point against which student learning can be measured. Identifying the extent of students' prior knowledge of a topic also enables teachers to plan subsequent learning experiences. This approach is known as evidence-based teaching.

Assessment techniques commonly used by teachers for diagnostic purposes include student observation, focused questioning and diagnostic tasks. The *First Steps in Mathematics* resources include many examples of mathematical tasks designed for the purposes of diagnostic evaluation.

Reading
The *First Steps in Mathematics* resources are available online at:

det.wa.edu.au/stepsresources

Activity

Use the *First Steps in Mathematics* materials to identify simple diagnostic tasks for Number, Measurement, Space and Chance. What does each diagnostic task tell you about students' *Key Understandings* in relation to the chosen topic?

Formative assessment

Formative assessment has come to be known as assessment *for* learning. Teachers provide students with formative feedback during instruction to guide (and in some cases, redirect) their learning. By providing feedback in response to student needs, teachers can help to maximise student learning.

There is a wide range of formative assessment techniques. A simple method that provides the teacher with instant feedback is to ask students to respond with simple gestures, such as holding a thumb up to indicate their understanding. A common variation on this technique is to provide students with a set of green, yellow and red cards (i.e., traffic lights)  which are held up to indicate their level of confidence and/or need for assistance.

It is also common for teachers to use questioning, quizzes, homework, observation, checklists, student presentations and student self-evaluation for the purposes of collecting formative data. It is important to understand during this phase that making mistakes is an accepted part of learning. Students who do not make any mistakes are unlikely to be working to their full potential. Remind students that the purpose of this type of assessment is to maximise their learning potential and not to make evaluative judgements. A key component of formative assessment is feedback from the teacher to the student that enables the student to move forward in their learning.

Summative assessment

Summative assessment is known as assessment *of* learning. Its purpose is to determine the extent of student learning at a *fixed* point in time—typically at the end of the teaching and learning cycle. Since student learning is not confined to particular teaching and learning cycles, summative assessment only provides a snapshot of students' knowledge and abilities at a specific point in time.

Teachers use a broad range of summative assessment strategies. Common techniques include:

- marking problems, exercises, assignments and student files
- collecting work samples
- standardised tests
- teacher-made tests
- interviews
- student presentations
- demonstrations
- projects, and
- learning portfolios.

Reporting

The purpose of reporting is to communicate accurate and useful information about student achievement in a concise manner. It should focus on what students can do and what they need to be able to do. It should provide information about what students need to do in order to move forward in their learning. This means that student progress in mathematics needs to be continually monitored and recorded.

A comprehensive assessment program typically provides a great deal more information than can be communicated in students' end of term reports. Teachers must therefore make evaluative judgements about student learning based on the range of assessment data that was gathered. These are sometimes referred to as *on-balance judgements*.

Activity

How would you evaluate each of the following report comments?
Give reasons for your answers.

1. *Alan is a very pleasant child who is extremely helpful in class. With more effort he is likely to achieve greater success in mathematics.*

2. *Brenda has developed a strong understanding of the fraction concept this term. She can model one-half and one-quarter in different ways. She is now working on extending her understanding to other unit fractions. Good job!*

3. *Chris spends a great deal of class time working on the computer and so often does not complete assigned tasks in mathematics. He does not seem to listen to instructions and so needs to learn to develop his attention span or his work will continue to suffer.*

4. *Denise is an extremely capable student. Although she usually finishes her work very quickly, she often makes careless errors. She needs to check her answers instead of disrupting others.*

5. *Edgar has a good grasp of combinations to ten. He now needs to focus on extending this idea to larger numbers. He is developing a good repertoire of mental calculation skills and is able to use a calculator quickly and efficiently. Well done!*

6. *Frank is a bit of a show-off in class. He likes to be the centre of attention and his mathematics has suffered as a result. I know you can do better!*

The following general advice may be useful when preparing written report comments:

- Think carefully about the information you need to convey.
- Be prepared well in advance of the reporting deadline.
- Be positive, focusing on what the student can do and what they need to do in order to make further progress.

- Focus on the student's learning and not on their personality traits.
- Write in a clear and professional manner.
- Carefully check spelling and grammar (or ask another teacher to do so).
- Review your comments a day or so later to see if you still agree.
- Be aware of school requirements.

Working collaboratively

The best educational outcomes can only be achieved when students, parents and teachers accept shared responsibility for student learning. Teachers should introduce themselves to parents as early as possible so they can begin to establish good working relationships. Many teachers send all parents an email early in the year to introduce themselves and establish their expectations. Other teachers maintain a class blog, send regular newsletters, etc.

Parents usually appreciate being informed about what is happening in the classroom. Most parents are interested in their child's education but are unsure what (if anything) they can do to help. It is much easier for teachers to address difficult issues with parents when they have already established a good working relationship. Be proactive; the first time that parents find out that a problem has occurred should *not* be when they receive their child's report!

The role of questioning

A discussion of classroom assessment techniques would be incomplete without considering the role that questioning plays in the learning environment. Using questions effectively is an essential skill that all teachers must develop. In fact, teaching students to ask effective questions is also a very important aspect of their learning. One way to develop your questioning skills is by observing an experienced teacher's use of questions in the classroom.

Activity

Observe your mentor teacher's use of questioning in the classroom. Think about the specific questions they chose to ask.

- How did they *direct* their questions?
- How did they *respond* to the answers provided by students?
- How long did they *pause* when waiting for students to respond?
- What did you notice about their *tone* of voice, their *phrasing* of questions and their *emphasis* of key words?

Teachers spend a great deal of time planning and asking questions and responding to answers. Teachers may ask literally hundreds of questions a day. The purpose of questions can be to gain feedback, focus discussions, arouse interest, encourage higher-order thinking, summarise major points, stimulate thinking and to evaluate learning.

Barry and King (1998) suggested that teachers should:

- Ask questions in an encouraging manner.
- Initially direct questions to the whole class. Then select one student to answer. This encourages all students to think about the answer.
- Redirect questions. Allow several students to respond to your questions, rather than accepting the first answer that is given. This leads to a greater degree of student involvement and increases the range of answers.
- Distribute questions. Make a point of distributing questions around the whole class.
- Wait before asking for an answer. Student responses are likely to be longer and more varied.
- Ask one question at a time. Many teachers are so anxious to get an answer that they repeat or rephrase the question before giving students a chance to respond.
- Avoid rhetorical questions as they preclude student involvement.

Teachers should also consider the way in which they respond to the answers given by students. It is recommended that teachers do not respond to random call-outs. Teachers should also acknowledge the contributions made by students, whether by verbal or non-verbal means. Experienced teachers can sometimes develop student answers and incorporate them into the lesson. This helps the students to feel that they are an important part of the learning process.

Sometimes teachers will need to assist students by providing hints or clues. If a class is having difficulty answering a question, sometimes the question can be reworded or changed so that it is clearer. If a response is partially correct, the teacher can assist by prompting for further information. Incorrect answers should be corrected, but it is important not to fall into the trap of answering your own questions (Barry & King, 1998).

Baroody and Coslick (1998) provided the following tips for asking effective questions in the mathematics classroom:

1. *Anticipate student thinking.* Teachers should think about the types of questions that are likely to arise as students attempt the mathematical learning task.
2. *Link to learning goals.* By asking questions that link to the curriculum, the teacher can help students to focus on key principles.
3. *Pose open questions.* Open questions encourage a variety of approaches and responses. Open questions intrinsically allow for differentiation since they allow students to respond at their own stage of development.
4. *Pose questions that actually need to be answered.* In addition to not asking rhetorical questions, teachers should ensure that the questions they ask contribute to student learning.
5. *Incorporate verbs that elicit higher levels of thought.* Verbs such as connect, elaborate, evaluate and justify require students to communicate their thinking and understanding.
6. *Pose questions that open up conversation.* The way in which teachers ask questions can lead to discussion of the big ideas in mathematics.

7. *Keep questions neutral.* Labelling questions as 'easy' or 'hard' can shut down the potential for student learning.
8. *Provide wait time.* The quality and quantity of answers will generally increase and less confident students are more likely respond.

In summary, this chapter has outlined the principles of assessment, described the different types of assessment and provided some guidance about the role of questioning in the primary mathematics classroom.

Chapter 9
Using Technology

Almost every aspect of teachers' work involves the use of technology in one form or another. For example:

- planning typically involves accessing curriculum documents and/or support materials (such as Scootle) online
- teaching may involve using an interactive whiteboard, computer, data projector, DVD player and/or iPad
- communicating with parents and/or colleagues will generally include the use of email and/or an online portal
- using the school's learning management system
- recording assessment data in a spreadsheet, and
- preparing student reports.

In many cases, teachers are also expected to overcome the kinds of minor technical issues that occur frequently in classrooms—such as connecting to a projector, poor signal strength and resetting forgotten passwords—often with little or no assistance. Where possible these issues should be addressed in the planning stage, either by rehearsing before the lesson or by having a backup plan in place. In this chapter, we will explore how emerging and existing technologies have changed, and will continue to change, aspects of teaching and discuss the implications for teachers.

Calculators

Calculators were one of the earliest forms of technology introduced to the primary classroom. Their introduction at this level was contentious and continues to be so. Critics feared that students would become over-reliant on calculators and fail to develop strong mental arithmetic skills. Unfortunately, these fears are realised if  teachers do not make judicious use of this technology. Calculators are not for everything but, when used wisely, they can support learning by assisting students to explore large numbers and number patterns. They are a useful tool to learn about place value and can be used to check answers when learning about computation.

Activity

Use a calculator to complete the following. Once you think you know what the pattern is, you can predict what the remaining answers will be.

143 × 7 =	143 × 28 =	143 × 49 =
143 × 14 =	143 × 35 =	143 × 56 =
143 × 21 =	143 × 42 =	143 × 63 =

What are the benefits of using a calculator for an activity such as this?

SAMR model of technology adoption

The speed at which emerging technologies (such as iPads) have been adopted by the public has exceeded almost all expectations. Approximately 300 million iPads have been sold since 2010. Although we recognise that touch technologies have the potential to radically change the nature of teaching and learning, we don't yet fully understand their long-term impacts on education. It is likely that they will continue to change education in ways that we can't yet even imagine. According to the SAMR model, teachers' use of technology in the classroom may be classified into four levels: Substitution, Augmentation, Modification and Redefinition. Each level is explored in more detail in the table below.

Level	Definition	Examples
Substitution	Technology acts as a direct substitute for existing tasks with no functional change.	Instead of a handwritten worksheet, students print a worksheet or complete it online.
Augmentation	Technology acts as a direct tool substitute with functional improvements.	Students take a quiz using a Google Form instead of using pencil and paper.
Modification	The use of technology allows an existing task to be redesigned with significant functional improvements.	Instead of writing an essay, an audio recording is made of the essay along with an original musical score.
Redefinition	Teachers use technology to create entirely new kinds of learning experiences that were previously impossible.	Students create an interactive eBook with student-created embedded media.

Planning with technology

In the past, millions of dollars were spent printing curriculum documents, distributing them to teachers and replacing them each time that the curriculum was updated. Making the Australian Curriculum available online has resulted in several advantages for teachers, including:

- instant access to the most up-to-date version of the curriculum
- content descriptions can be cut and pasted into programs and lesson plans
- the ability to access the curriculum virtually anywhere, and
- parents have access to the curriculum.

With the release of the Australian Curriculum, for the first time it was possible for content descriptions to be directly linked to teaching resources. Previously teachers would have to find resources that matched to specific content descriptions. Now teachers can select appropriate resources from the variety of resources available through Scootle (see Chapter 7).

Communicating with technology

The ability to communicate clearly and effectively is a prerequisite for successful teaching. Teachers can communicate information to parents in a very time-efficient manner using communication technologies such as blogs and email. Since the technology also serves to document their communication, it is important that teachers think carefully about the message they wish to communicate.

Teaching commitments mean that there are times when teachers are unable to take phone calls. Asynchronous forms of communication (such as email) allow teachers to address their correspondence when they are in a position to give it their undivided attention. Attaching files or hyperlinks to emails also allows teachers to share information with parents or students in an efficient manner, such as when there is a need to send work home in the case of absence.

Collaborating with technology

Widespread use is now made of cloud-based technology. Many companies (such as Dropbox) provide free access to cloud-based storage. Other companies (such as Google) provide access to cloud-based applications that allow multiple authors to collaborate on a single document. This technology has numerous applications for teachers, such 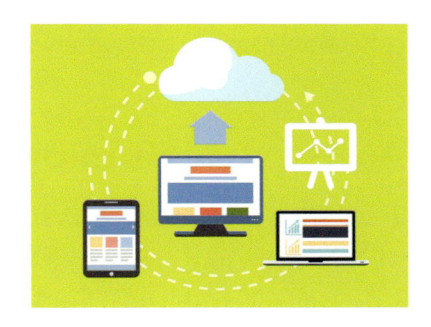 as the ability to collaborate with other teachers when designing a teaching program, team teaching and working with education assistants.

In the classroom, students can use cloud-based technology to work on group projects. Students can also collaborate with students from other schools or other countries. Cloud-based storage can also protect work from loss due to theft or fire, although teachers and students must still consider issues surrounding data security and access.

Teaching resources

Modern applications allow teaching resources to be created quickly and efficiently using a wide variety of built-in templates. Examples of the free templates available in Microsoft Word include merit

certificates, lesson plans, flash cards, weekly task sheets, assignments and note taking sheets. Resources purchased online can often be accessed instantly. Remember that purchasing teaching resources will in many cases be tax deductible for a teacher.

For teachers, there are many advantages of storing resources in digital form. Once a resource has been created in digital form, it can be readily modified to suit the learning needs of particular students and/or classes. Digital storage is highly space efficient and very affordable, meaning that backup copies can be made at minimal cost. It is important for teachers to have an organised resource collection to reduce the amount of time wasted looking for documents.

Assistive technologies

Touch technologies (such as iPads) have proven to be invaluable for students with sensory impairments. Many of these devices have built-in features to assist students with visual impairments (such as resizable text and high-contrast modes). *Dragon Dictation* is a mobile application (i.e., app) that converts speech into text. Other apps can be used to convert text to speech, recognise handwriting and assist with speech therapy.

There are also many apps which support the development of early numeracy. These have not been reviewed here as this information is rapidly superseded. Schools that make use of class sets of iPads often have a Standard Operating Environment that may limit the selection of apps that can be used.

Assessing with technology

The multimedia capabilities of modern smart phones and tablets allow teachers to incorporate a much broader range of assessment techniques into their teaching and learning programs. Data can be gathered and stored in digital portfolios for later retrieval. Generally, files are automatically tagged with properties such as the date and time they were created which helps to provide a record of student progress. Students should be encouraged to take part in the selection of work samples to include in their portfolios.

The kinds of assessment products that can be created with tablets and smart phones include, but are certainly not limited to:
- digital storybooks
- stop motion animation/Slowmation (www.slowmation.com)
- audio and video podcasts
- time lapse photography, and
- interactive quizzes.

Reporting with technology

Teachers can spend a great deal of time preparing written report comments. Report writing often happens during particularly busy times of the year and must be completed within tight deadlines. Schools often use purpose-built software for reporting and software for preparing reports. It is recommended that you draft your written comments in Microsoft Word (or similar) before pasting them into the reporting software. This serves a number of purposes:
- You can use Word to check your spelling and grammar
- Word is more familiar and easier to use than most reporting software, and

- Reporting software sometimes lacks features such as being able to change the size of the text box or font, which makes comments hard to read and edit.

Flipping the classroom

Salman Khan is the founder of Khan Academy, a not-for-profit educational organisation whose mission is to provide 'a free, world-class education for anyone, anywhere' (Khan Academy, 2017). The Khan Academy site provides free access to thousands of online instructional videos covering topics in mathematics, science, history, economics, computer coding and many other areas. The Khan Academy website directs students through the work they need to do in order to move forward in their learning and helps teachers and parents keep track of their progress. Teachers can also sign up for a free account that they can use with their classes.

In his 2011 TED Talk (available at www.ted.com), Salman Khan proposed an alternative to the traditional teacher-focused approach. Khan argued that the lecture style approach used in many classrooms fails to cater to the needs of individual students. Instead, he suggested teachers free up time in the classroom by assigning online lecture-style content for students to work through at home. This approach has become known as *flipping the classroom.*

Flipping the classroom has potential advantages for both students and teachers. Students can work at their own pace and can move on to more challenging work when they are ready. The teacher can also make better use of limited classroom instructional time by spending less time giving one-size-fits-all lessons and more time working with individual students. Khan argues that using technology in this manner allows teachers to humanise the classroom by spending more quality instructional time with their students. This approach may be more suitable with certain age groups and content areas within mathematics.

Khan Academy is a not-for-profit education platform for learning mathematics, science, economics, history, computer coding and many other subjects. Sign up for a free account at www.khanacademy.org

Reading
You can read more about Salman Khan's approach in his book:

Khan, S. (2012). *The one world school house: Education reimagined.* New York: Hachette Book Group.

Questions to consider

1. How do the roles of the teacher and the student change in the flipped classroom?
2. What role does technology play in the flipped classroom?
3. What challenges are likely to be encountered if using this approach?

Teaching with technology

There is a wide range of resources available online for the teaching and learning of mathematics. Some of the most well-known and useful are briefly reviewed below. Many of these are provided by teacher professional organisations such as the Australian Association of Mathematics Teachers (AAMT) and the National Council of Teachers of Mathematics (NCTM) in the United States.

The Nrich site is provided by Cambridge University and Youcubed is provided by Stanford University. Care should be taken when searching for resources outside of Australia, since the indicative year levels may not match those of the Australian Curriculum.

Top Drawer Teachers (AAMT)

https://topdrawer.aamt.edu.au/

The Top Drawer Teachers site has a collection of resources for teaching fractions, mental computation, patterns, reasoning and statistics. Each collection includes discussion of the big ideas, common student misunderstandings, examples of good teaching and a large range of ready-to-use learning activities.

Youcubed at Stanford University

www.youcubed.org

Professor Jo Boaler's site provides many short videos on how the brain learns mathematics and how students can adopt a growth mindset to learning mathematics. The site also provides suggestions for teachers and a wide range of ready-to-use mathematical learning activities.

Nrich: The home of rich mathematics

nrich.maths.org

For 20 years, the Nrich Project at Cambridge University has provided resources for teachers and students of all year levels and provided professional development for teachers. The Nrich site contains an extensive range of resources which can be searched by topic.

Illuminations (NCTM)

illuminations.nctm.org

The Illuminations site provides a broad range of resources for teaching and learning mathematics, including interactive tools for students and instructional support for teachers. The site contains over 700 lesson plans and over

100 virtual manipulatives, applets and games (including seven mobile apps for iOS and Android).

GeoGebra

GeoGebra is free geometry and algebra software available for across almost all platforms. It also runs from within a browser, meaning that it does not need to be installed on school computers (which in some circumstances may not be permitted). GeoGebra provides free access to many features that would otherwise require an expensive graphics calculator.

Students and teachers can use GeoGebra to create diagrams, demonstrations, activities, games and workbooks, which can be then shared with others. Just under one million resources have been created and shared and these are available to be downloaded using the Materials tab (i.e., www.geogebra.org/materials). These include many excellent lessons and learning activities.

 GeoGebra may be downloaded for Mac, PC, Google ChromeOS, Linux, iOS and Android from www.geogebra.org/download

Purchasing software

Schools and universities generally purchase site licences for software packages that their staff and students need to do their work (such as Microsoft Office). You should be aware that many companies offer significant discounts to students and staff who purchase software for educational purposes. Such educational discounts are offered on the understanding that these products are not resold or used for commercial use.

 Microsoft Office 365 University is a subscription package available to eligible students and university staff. You can check the cost and what is available in the package here: www.office.com/getoffice365

This chapter has discussed the use of technology in the mathematics classroom, including the SAMR model of technology adoption and how teachers make use of technology in the classroom and other aspects of their professional work.

Part B
mathematical knowledge

Chapter 10

Whole Numbers

For thousands of years humans have used numbers to quantify aspects of the world. Over time many different number systems have been used, each designed to address the needs of the society that they served. The Hindu-Arabic number system that we use today was originally invented by Indian mathematicians between the 1st and 4th centuries AD. Around the 8th century, the system was adopted by Persian and Arabic mathematicians, who called them 'Hindu numerals'. Later they came to be called 'Arabic numerals' when they were introduced to Europe by Arabian merchants.

The Base 10 number system

The Hindu-Arabic numerals are based on a set of 10 symbols: 1, 2, 3, 4, 5, 6, 7, 8, 9 and 0. In Europe, the system was popularised by Italian mathematician

Leonardo Fibonacci in 1202, who wrote that with just 'these nine figures, and with this sign 0 . . . any number can be written'. Born around 1175 in Pisa, young Leonardo travelled with his father (a wealthy merchant) and first encountered the Hindu-Arabic system in northern Africa.

Fibonacci believed that the Hindu-Arabic system had many advantages for European merchants and mathematicians over the Roman numerals that were in use at the time. For example, 1 989 expressed using Roman numerals is MCMLXXXIX, 1 999 is MCMXCIX and 2 000 is MM. Performing calculations with Roman numerals is even more tricky. The Hindu-Arabic system introduced the idea of *place value*, which made calculations much easier. Fibonacci wrote one of the first mathematics text books which he called *Liber Abacci* (meaning the book of the abacus). It was completed in 1202 and as well as showing how to perform calculations using the new system, this book contained some problems to be solved including the famous rabbit problem, used to illustrate the number pattern which still bears his name, the Fibonacci sequence.

The Base 10 system is an example of a *positional number system*. This means that the place of a digit within a number affects its value. For example, while 45 and 54 consist of the same digits, in one case the number represented consists of *4 tens* and *5 ones* while in the other it consists of *5 tens* and *4 ones*. The concept of place value needs to be carefully developed over the primary years as it is one that many students find confusing.

Reading, writing and saying numbers

In the Base 10 system, each set of three digits represents the number of hundreds, tens and ones in the corresponding place value grouping. For example, the number represented below is twelve *million*, five hundred and sixty-seven *thousand*, four hundred and ninety-one.

Hundreds	Tens	Ones	Hundreds	Tens	Ones	Hundreds	Tens	Ones
Millions			Thousands			Ones		
	1	2	5	6	7	4	9	1

This pattern extends to larger place value groupings (i.e., billions, trillions, etc.). In order to make large numbers easier to read on sight, the accepted convention is to leave a space between each set of three digits e.g., 12 567 491. (An earlier convention was to use commas for the same purposes. This is still common practice in some countries.)

Expanded form

The Base 10 system provides an efficient way of representing numbers. Using expanded notation, we can split numbers to show their component parts. For example, 432 can be thought of as 4 hundreds, plus 3 tens, plus 2 ones, which is represented symbolically as shown below:

$$432 = (4 \times 100) + (3 \times 10) + (2 \times 1)$$
$$= 400 + 30 + 2$$

Comparing numbers

In order to compare numbers in the Base 10 system, we first look at the left hand digit, which is the digit with the largest place value in the number. One number is larger than another if its left hand digit has a greater value than the left hand digit of the other number. The *greater than* symbol (>) is used to show that the number on the left is larger than the number on the right, while the *less than* symbol (<) shows that the number on the left is less than the number on the right.

In the example below, the left hand digits are in the **thousands** column. If the digits in the thousands column are the same, then we move to the next largest place value column (i.e., **hundreds**). In the example below, 2 121 and 2 112 are both larger than 2 013. But since both 2 121 and 2 112 have the same number of hundreds, we must compare them by looking at the digits in the **tens** column. This tells us that 2 121 is a larger number than 2 112 (i.e., 2 121 > 2 112). This same idea extends to comparing larger numbers and decimal numbers.

Thousands	Hundreds	Tens	Ones
2	0	1	3
2	1	2	1
2	1	1	2

Using expanded notation, these numbers can be expressed as:

$$2000 + 0 + 10 + 3$$
$$2000 + 100 + 20 + 1$$
$$2000 + 100 + 10 + 2$$

Index notation

Just as multiplication provides a more concise way of representing repeated addition, index notation is a kind of shorthand for repeated multiplication. The use of index notation (or **indices**) involves a number, called the **base**, that has been raised to a power (or **exponent**). To evaluate an index number, we multiply the base number by itself as many times as indicated by the power. For example, we can write 8×8 as 8^2 and say *'eight to the power of two'* or simply *'eight squared.'* Similarly, we can write $10 \times 10 \times 10$ as 10^3 and say *'ten to the power of three'* or simply, *'ten cubed.'*

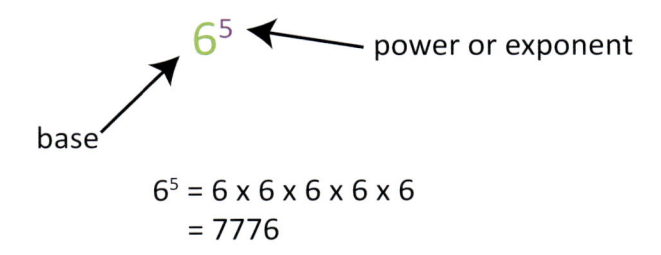

$$6^5 = 6 \times 6 \times 6 \times 6 \times 6$$
$$= 7776$$

Scientific notation

Scientific notation is a convenient way of writing very large or very small numbers. A number written in scientific notation is said to be in *standard form*. In the Base 10 system, multiplying and dividing by powers of 10 has the effect of changing the place value of a digit. For example, since 2 000 can be written as $2 \times 10 \times 10 \times 10$ we can use scientific notation to write 2 000 as 2×10^3.

A number written in scientific notation is expressed as a decimal number with just one digit to the left of the decimal point multiplied by a power of 10. For example, 6.28×10^3 represents 6 280. Notice that the effect of multiplying by 10^3 is to move the digit that was in the ones place (i.e., 6) to the thousands place.

Scientific notation also provides a convenient way of expressing small numbers. For example, 4.91×10^{-2} represents 0.0491. Notice that the effect of multiplying by 10^{-2} moves the number that was in the units place to the hundredths place. (In other words, multiplying by 10^{-2} has the same effect as dividing by 100.)

Rounding

Rounding is the process of reducing the amount of non-zero digits in a number but keeping the value approximately the same. When rounding to a particular accuracy (for example, the nearest *ten*), we must decide whether the number is more than half-way to the next division. For example, 236 is more than half-way from 230 to 240 and so we round up. Some everyday examples of rounding include:

- the distance from home to school is rounded to the nearest kilometre
- the amount charged for shopping is rounded to the nearest 5 cents, and
- rounding the amount of time a task will take to the nearest hour.

Original value	Required accuracy	Approximate value	Accuracy
236	Nearest 10	240	98.3%
9 451	Nearest 100	9 500	99.5%
14 385	Nearest 1 000	14 000	97.3%

A number line can be helpful in deciding which division the number is closest to. The number line below shows 14 385 being rounded to the nearest thousand. As can be seen, 14 385 is closer to 14 000 than it is to 15 000.

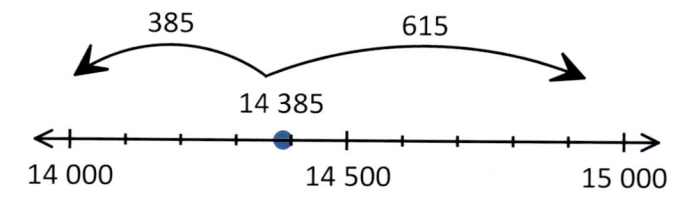

In cases where the number to be rounded is exactly half-way between the two divisions (for example, 150 is half-way between 100 and 200), the accepted convention is to always round up (i.e., 150 is rounded up to 200).

Exercise 10.1

1. Write out the following numbers using words:
   a. 6 023
   b. 12 357
   c. 144 057
   d. 1 050 060
2. Write out the following numbers using numerals:
   a. Four hundred and twenty-three thousand and sixty-four
   b. Eight thousand nine hundred and forty-seven
   c. One thousand, one hundred and eleven
   d. Nineteen thousand and fourteen
3. Express each of the following numbers using scientific notation:
   a. 625 000
   b. 0.00045
   c. 41 200
   d. 1 250 000
4. Write out the following numbers using numerals:
   a. 4.02×10^5
   b. 6.714×10^{-2}
   c. 9.81×10^1
   d. 3.1×10^0
5. Round each of the following numbers to the required accuracy:
   a. 14 799 to the nearest ten
   b. 2 010 to the nearest hundred
   c. 401 to the nearest thousand

Addition

We *add* or *combine* to find the *sum* or *total*. We also use addition to *increase* an amount by a fixed quantity. For example we might say:

- the *sum* of 7 and 8 is 15
- 7 *plus* 8 equals 15

The vocabulary of addition

When 75 and 48 are added, the 75 and 48 are described as *addends* and the result is referred to as the *sum*.

Addend	→		7	5
Addend	→	+	4	8
Sum	→	1	2	3

In the written algorithm (an algorithm is the step-by-step process for carrying out a written calculation) for addition, two (or more) numbers are aligned according to their place values. Calculation proceeds from right to left, adding the digits in the ones column first. We add the digits and record the total. If the total exceeds 9, then we record the ones digit in the answer and add the tens digit to the digits in the tens column. We then proceed to add the digits in the tens column and continue with this process until all columns are added.

Example

In order to add 38 and 45, we first align the place value columns. We begin from the right, adding 8 ones to 5 ones to get a total of 13 ones. We exchange 10 of these ones for 1 ten in the tens column, ensuring that we record the exchange as shown below. We then add 3 tens to 4 tens, remembering to add the additional ten, to get a total of 8 tens. This gives the final answer of 83.

$$\begin{array}{r} {}^{1}3 \quad 8 \\ +\ 4 \quad 5 \\ \hline 8 \quad 3 \end{array}$$

In the classroom, the use of any standard algorithm should only be introduced as an extension of students' existing mental and informal written strategies. Emphasising the written algorithm too early can be counter-productive and lead to students losing sight of the significance of place value and failing to develop the ability to judge the reasonableness of their answers.

Questions to consider

1. Why don't we work from left to right when using the written algorithm?
2. Does it make a difference if we write the addends in a different order?
3. What aspect(s) of the written algorithm for addition might have the potential to cause confusion?

Exercise 10.2

Work through the following examples of the written addition algorithm. No calculators!

a.
```
    7 8
+   3 9
───────
```

b.
```
  2 7 4
+ 3 1 6
───────
```

c.
```
    9 6 7
+ 1 0 8 9
─────────
```

d.
```
    1 2
    3 4
  1 2 7
+ 2 0 9
───────
```

e.
```
    7 4
    8 9
  1 0 6
+   1 5
───────
```

f.
```
      8 7
    1 1 9
    2 1 6
+ 2 4 9 1
─────────
```

Mental strategies for addition include:
- Adding from the left (sometimes called *Front Loading*)
- Bridging Tens
- Counting Forward (or using a number line), and
- Compensation

In the Front Loading approach, calculation proceeds from left to right. For example, to add 427 and 235:

$$427 + 235 = (400 + 200) + (20 + 30) + (7 + 5)$$
$$= 600 \quad + \quad 50 \quad + \quad 12$$
$$= 662$$

By performing the calculation in this order, the largest numbers are dealt with first (i.e., 400 + 200 = 600) giving a better approximation of the answer than if we had started with the ones (i.e., 7 + 5 = 12). As a mental strategy, it is usually easier to add hundreds and tens before adding the ones.

The Bridging Tens approach works by partitioning or rearranging numbers to make use of convenient groupings of 10, 100 or 1 000. For example:

$$9 + 4 = 9 + (1 + 3)$$
$$= (9 + 1) + 3$$
$$= 10 + 3$$
$$= 13$$

$$8 + 7 + 2 = (8 + 2) + 7$$
$$= 10 + 7$$
$$= 17$$

$$68 + 45 = 68 + (32 + 13)$$
$$= (68 + 32) + 13$$
$$= 100 + 13$$
$$= 113$$

$$68 + 27 + 12 = (68 + 12) + 27$$
$$= 80 + 27$$
$$= 107$$

The Counting Forward approach involves starting with the larger number and counting forwards the required number of steps on the number line. For example, to calculate 27 + 14 by counting on, we would begin at 27 and step forward 14. This can be done as 14 steps of one unit or by skip counting by 10 and then 4.

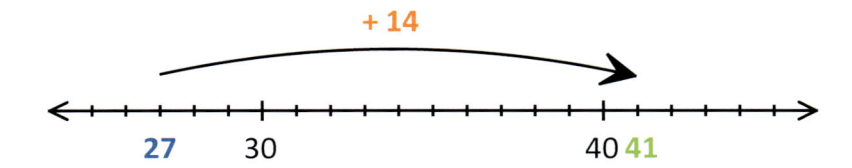

The Compensation approach involves adding more than is actually necessary and then adjusting (or compensating) the result. For example, to add 88 and 26, we could begin with 88 and add 30 to get 118. Since we have added 4 more than was required, we subtract 4 to get 114. This can also be illustrated using a number line:

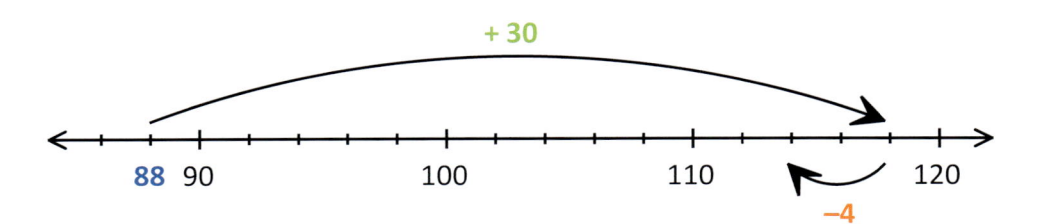

Subtraction

Subtraction is used to find the *difference* between two quantities or to *decrease* an amount by a fixed quantity. For example we might say:

- the *difference* between 12 and 19 is 7
- 23 *take away* 8 equals 15

The vocabulary of subtraction

When 48 is subtracted from 75, the 75 is called the *minuend* and the 48 is called the *subtrahend*. The answer, 27, is known as the *difference*.

Minuend	→	7	5
Subtrahend	→	− 4	8
Difference	→	2	7

In the written algorithm for subtraction, we align the two numbers according to their place values, with the larger number on top. Beginning with the right hand column (i.e., ones), subtract the bottom number from the top number. (This will require exchanging or regrouping if the bottom number is smaller than the top number.) Record the single-digit answer and move to the next column to the left.

Example

In order to subtract 149 from 267, we first align the place value columns. We begin from the right and recognise that we cannot subtract 9 ones from 7 ones without regrouping. To do this, we exchange one of the 6 tens in the top number for 10 ones. Now we subtract 9 from 17, giving 8 in the ones column. Moving to the tens column, we subtract 4 tens from the 5 remaining tens, giving 1. Subtracting 100 from 200 gives 1 in the hundreds column and our final answer of 118 as shown below.

$$2 \quad {}^{5}\!\!\not{6} \quad {}^{1}7$$
$$- \quad 1 \quad 4 \quad 9$$
$$\overline{\quad 1 \quad 1 \quad 8 \quad}$$

Exercise 10.3

Work through the following examples of the subtraction algorithm. No calculators!

a.
```
    7  1
 -  2  9
 _____
```

b.
```
    3  4  1
 -  2  8  6
 _____
```

c.
```
    2  9  3  7
 -  1  0  8  9
 _____
```

d.
```
    3  0  0
 -  2  9  1
 _____
```

e.
```
    7  0  6
 -  3  4  1
 _____
```

f.
```
    2  2  1  0
 -  1  0  9  1
 _____
```

Questions to consider

1. Why don't we work from left to right when using the subtraction algorithm?
2. Does it make a difference if we write the smaller number on top?
3. What aspect(s) of the written algorithm for subtraction might cause confusion?

Mental strategies for subtraction include:
- Counting Back (or using a number line),
- Counting Forward, and
- Adding a Constant.

Counting Back is simply the opposite of counting forward. Start with the larger number and count back the required number of steps. For example, to calculate 65 – 17, we can count back 17 to get 48. This could be broken down as take away 15 then take away 2.

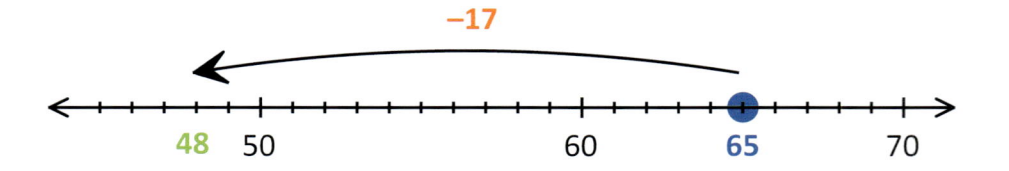

Alternatively, we can calculate 65 − 17 by first subtracting 20 and then compensating by adding 3.

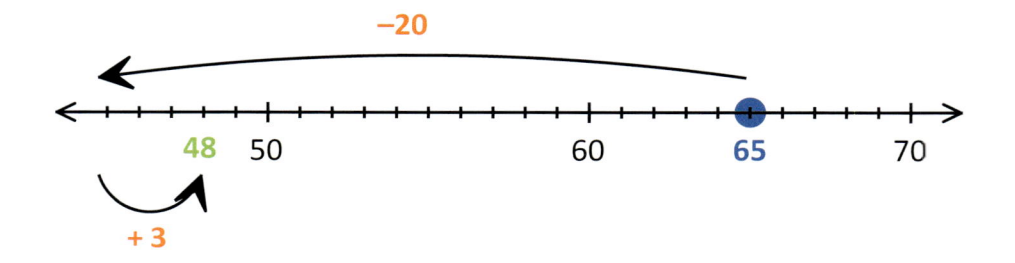

Another strategy is to Count Forward from the smaller number. For example, to calculate the difference between 65 and 17, we could add 3 to 17 to reach 20, then add 40 to 20 to get 60 and finally add 5 to 60 to reach 65. The total difference is then 40 + 3 + 5 = 48.

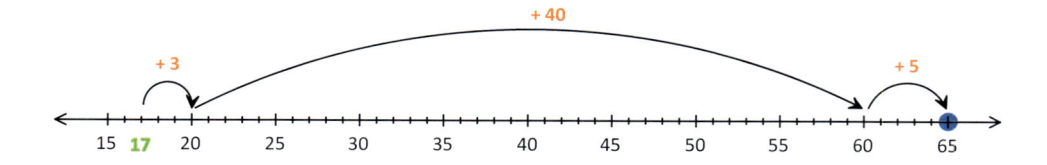

We could also Add a Constant to both numbers to simplify the subtraction. Instead of calculating 65 − 17, we could calculate (65 + 3) − (17 + 3) = 68 − 20. In both cases, the difference is 48. Using a number line, it can be shown that the difference between 65 and 17 is the same as the difference between 68 and 20. (Check that you agree with this!)

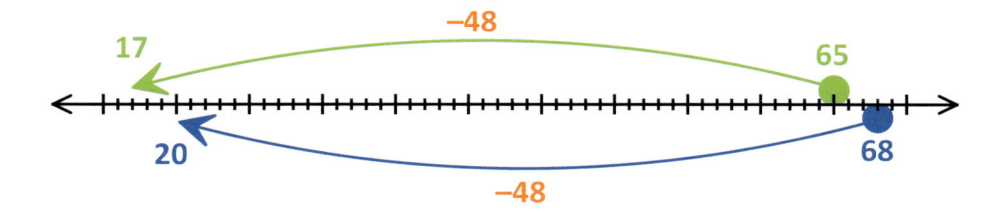

Multiplication

Multiplication is the process of repeated addition. When we multiply two or more numbers, the answer is called the *product*. The numbers being multiplied are called *factors*. When multiplying we might say:

- The *product* of 6 and 7 is 42
- 7 *times* 9 equals 63
- 9 *sets of* 4 are 36
- 4 *lots of* 6 are 24

The vocabulary of multiplication

When 75 is multiplied by 8, we describe 75 as the *multiplicand* and 8 as the *multiplier*. The result, 600, is called the *product*.

Multiplicand	→		7	5
Multiplier	→	x		8
Product	→	6	0	0

The written algorithm for multiplication uses the *distributive property* of multiplication over addition, which states that $a \times (b + c) = a \times b + a \times c$. First the two factors are aligned according to their place values. The factor with the most digits is generally placed on top (although this is not a requirement). Calculation proceeds from right to left. Multiply each of the digits of the top number (beginning with the ones digit and proceeding to the left) by the ones digit of the bottom number, writing down the ones digit of the product and carrying where necessary.

We then turn our attention to the tens digit of the bottom number. Since we are now multiplying by the tens digit, we place a zero as a placeholder in the ones column underneath our previous calculations. We then multiply each of the digits of the top number (beginning with the ones digit and proceeding to the left) by the tens digit of the bottom number, in each case writing down the ones digit of the product and carrying where necessary.

If necessary, we then turn our attention to the hundreds digit of the bottom number. Since we are now multiplying by the hundreds digit, we place zeroes as placeholders in both the ones and tens column underneath our previous calculations. We then multiply each of the digits of the top number (beginning with the ones digit and proceeding to the left) by the hundreds digit of the bottom number, in each case writing down the ones digit of the product and carrying where necessary.

We continue in this manner until we have calculated all of the partial products. The final step is to add these partial products, taking care to ensure that the place values are aligned correctly, to get the final answer.

Example

Using the written algorithm to calculate 432×14.

		¹4	3	2
	x		1	4
This row shows 432×4 →	¹1	7	2	8
This row shows 432×10 →	4	3	2	0
This row shows 432×14 →	6	0	4	8

Exercise 10.4

Work through the following examples of the standard written algorithm for multiplication. No calculators!

a.   2   8   b.   7   4   c.   6   1   8
  x   3   1   x   1   6   x   3   4
 _____   _____   _____

 _____   _____   _____

 _____   _____   _____

Questions to consider

1. Why don't we work from left to right when using the written algorithm?
2. Does it make a difference if we switch the top and bottom numbers in the multiplication?
3. Does it make a difference if we first multiply each digit of the top number by the tens digit of the bottom number first, and then by the units digit second?
4. What aspect(s) of the written algorithm for subtraction might have the potential to cause confusion?

Mental strategies for multiplication include:
- Doubling
- Doubling and Halving
- Factoring, and
- Using the Distributive Property.

Doubling is a simple mental strategy for multiplying by 2, 4 or 8. For example, we can double each digit in 123 to get 246. To double 246 we can double each digit as long as we pay attention to place value: $2 \times 246 = 2 \times 200 + 2 \times 40 + 2 \times 6$. Therefore $2 \times 246 = 400 + 80 + 12 = 492$.

Doubling and Halving works on the principle that multiplication and division are inverse operations. To calculate 25×16, we can double 25 and halve 16 to get 50×8. We can repeat the process, doubling 50 and halving 8 to get 100×4. All of these products have the same answer i.e., $25 \times 16 = 50 \times 8 = 100 \times 4$. Factoring works along similar lines; sometimes rearranging the order of the factors makes the product easier to calculate. For example, 20×36 can be written as $(2 \times 10) \times (6 \times 6)$.

The Distributive Property can be used to partition one or both factors in the product. For example, to calculate 397×12, we can calculate 400×12 and then subtract 3×12. This is because $397 \times 12 = (400 - 3) \times 12$. If we partition both factors, for example $25 \times 18 = (20 + 5) \times (10 + 8)$, we must calculate four partial products i.e., $(20 \times 10) + (5 \times 10) + (20 \times 8) + (5 \times 8)$. This is best illustrated using grid multiplication:

	20	5
10	20 x 10 = 200	5 x 10 = 50
8	20 x 8 = 160	5 x 8 = 40

Division

Division is the process of repeated subtraction. When one number is divided by another, we are finding out how many times the second number can be subtracted from the first. When dividing 24 by 6 we might say:

- 24 *divided by* 6 is 4
- 24 *shared between* 6 is 4
- 6 *goes into* 24 four times
- The *quotient* of 24 and 6 is 4

The vocabulary of division

The division algorithm gives the *quotient* of two numbers. The *divisor* (8) is placed to the left of the *dividend* (600). The result of the calculation is the quotient (75).

Quotient →   7   5

Divisor   →   8 | 6   0   0   ← Dividend

For the example 1 398 ÷ 6, the first step is to determine how many times the divisor (6) divides into the left hand digit of the dividend and write this number above the first digit of the dividend. In this case, 6

$$2 \quad 3 \quad 3$$
$$6 \overline{)1 \quad 3 \quad {}^19 \quad {}^18}$$

divides into 13 twice, with a remainder of 1. The remainder is written as a carry digit above and to the left of the next digit in the dividend. This process is repeated for each digit in the dividend.

If a number divides into the dividend (i.e., the amount being divided) leaving no remainder, the divisor is called a factor of the dividend. This only occurs when the dividend is a multiple of the divisor. A clear understanding of factors and multiples is essential when working with fractions. If there is a remainder, as occurs when $453 \div 2$, it can be expressed as a fraction (e.g., $226\frac{1}{2}$), decimal (e.g., 226.5) or left as a remainder (e.g., 226 r1).

Exercise 10.5

Work through the following division problems. No calculators!

a. $7\overline{)11389}$ b. $4\overline{)1174}$ c. $3\overline{)12345}$

Questions to consider

1. Why don't we work from right to left when using the written algorithm for division?
2. Does it make a difference if we switch the divisor and dividend in a division problem?
3. What aspect(s) of the written algorithm for division have the potential to cause confusion?

Order of operations

The *Order of Operations* or *Rule of Order* is the agreed-upon convention used when performing mathematical calculations. The order of operations ensures that there is a unique answer to any calculation. Operations are performed in a specific order as we move from Left to Right through a calculation, with highest priority given to operations in Brackets, then Indices, Multiplication and Division, and Addition and Subtraction (BIMDAS).

Note: Multiplication and division have equal priority, as do addition and subtraction. These operations are performed in the order they occur in the calculation, so if division occurs before multiplication it is done first.

In the examples below, red has been used to indicate the part of the calculation that is done first.

Examples

1. $15 + (6 - 2) \times 3$
   $= 15 + 4 \times 3$
   $= 15 + 12$
   $= 27$

2. $20 \div 2^2 + 7$
   $= 20 \div 4 + 7$
   $= 5 + 7$
   $= 12$

3. $10 \times (5 + 2 \times 4)$
   $= 10 \times (5 + 2 \times 4)$
   $= 10 \times (5 + 8)$
   $= 10 \times 13$
   $= 130$

4. $8 \div 4 \times 2$
   $= 2 \times 2$
   $= 4$

Exercise 10.6

Use the order of operations to perform the following calculations.

1. $6 \times 7 - (4 + 4) + 3$

2. $(2^2 - 1) \times (3 + 1)$

3. $2 \times (4 - 2)^2 + 3$

4. $8 + (4 \div 2) \times 6 - 1$

5. $12 \div 3 \times (15 - 6) + 3$

6. $(12 - 5) + 9 \times 3$

7. $3 \times \dfrac{(4 - 2)^2}{2} - 1$

8. $\dfrac{8 + 4 \div 2}{5} + 6 \times 2$

9. $4 - 3 + 2 \times 5 - 4 \div 4$

10. $20 \div (4 - (10 - 8))^2$

Chapter 11

Fractions

Many people consider fractions to be one of the most challenging aspects of mathematics. It is not uncommon for people to memorise rules for performing fraction calculations with little or no understanding of the processes involved. In this chapter, we revise the fundamental concepts of fractions, the many different ways in which fractions can be represented and explore the reasons why fraction calculations are performed as they are.

The fraction concept

The simplest example of a fraction is where a whole is divided into two halves. Mathematically, we can write this as $1 \div 2$ or $\frac{1}{2}$ or 0.5. We can also think of practical examples, such as a sandwich that has been divided into two equal parts. In order to make sense of the fraction (i.e., $\frac{1}{2}$) we must first understand what the whole represents. The whole can be a number, an object or a collection. For example, we can have half of 60, half an orange or half a box of crayons.

A proper fraction consists of a top number (i.e., the **numerator**) that is smaller than the bottom number (i.e., the **denominator**). The denominator shows the number of equal parts into which the whole has been divided and the numerator shows how many parts we have. The denominator also serves to name the fraction. In the example shown here, the denominator gives the fraction the name 'quarters' (also called 'fourths' in the USA).

$$\frac{3}{4}$$

The bar that separates the numerator and denominator is called a vinculum. Mathematically, the fraction bar is equivalent to a division sign. That means that three-quarters $\left(\frac{3}{4}\right)$ is exactly the same as $3 \div 4$, which can also be written as 0.75 or $\frac{1}{4} + \frac{1}{4} + \frac{1}{4}$.

It is also possible to have fractions in which the numerator is larger than the denominator. These are known as **improper fractions**. Improper fractions are fractions whose value is larger than one whole. For example, $\frac{5}{4} = 1\frac{1}{4}$.

$$\frac{5}{4}$$

Improper fractions can be written as mixed numbers by separating them into the whole number part and the fractional part. In the example shown, this involves recognising that four-quarters (i.e., $\frac{4}{4}$) is equal to a whole. That is:

$$\frac{5}{4} = \frac{4}{4} + \frac{1}{4} = 1\frac{1}{4}$$

<u>**Note:**</u> It is common for students to think of fractions as only numbers that are smaller than a whole. Some students also confuse fractions with numbers that are less than zero (i.e., negative numbers). This is not the case.

Equivalent fractions

Fractions are **equivalent** if they represent the same numerical quantity. Equivalent fractions can be calculated by multiplying both the numerator and the denominator by the same non-zero number. This is sometimes known as renaming the fraction.

Note: An infinite number of equivalent fractions exist for any given fraction. For example, fractions equivalent to $\frac{1}{2}$ include $\frac{2}{4}, \frac{3}{6}$ and so on.

Activity

Write down two fractions that are equivalent to:

a. $\frac{5}{8}$

b. $\frac{3}{4}$

c. $\frac{5}{4}$

d. $\frac{3}{7}$

Comparing fractions

Fractions are most easily compared if they have the same denominator. Fractions which have the same denominator are known as like fractions. If this is the case, the fractions are made up of equal-sized pieces. The fractions may then be compared by looking at the numerators. For like fractions, the fraction with the largest numerator represents the largest quantity.

Fractions with different denominators are known as unlike fractions. These fractions are made up of different-sized pieces making them harder to compare. For example, it is difficult to say whether $\frac{4}{5}$ is greater than $\frac{7}{11}$ without further calculation. In these cases, it is useful to rename the fractions using a common denominator to enable quick comparison.

Example

To compare $\frac{3}{4}$ and $\frac{5}{6}$, we could rename $\frac{3}{4}$ as $\frac{9}{12}$ and $\frac{5}{6}$ as $\frac{10}{12}$. Instead of trying to compare quarters and sixths, we rename the fractions by finding two equivalent fractions that share the same denominator. This process creates like fractions (i.e., fractions that have the same size pieces) which are easy to compare. Clearly $\frac{10}{12}$ is larger than $\frac{9}{12}$ and so $\frac{5}{6}$ is larger than $\frac{3}{4}$.

Note: A simple way to find a common denominator is to multiply the original denominators together, although this does not always produce the lowest common denominator e.g., multiplying the denominators of $\frac{1}{4} + \frac{1}{6}$ gives a common denominator of 24, but the *lowest common denominator* is 12 and this is an easier number to work with.

Activity

Choose the larger fraction of each pair by finding a common denominator:

a. $\frac{3}{8}$ and $\frac{5}{6}$ b. $\frac{6}{7}$ and $\frac{2}{3}$

c. $\frac{4}{5}$ and $\frac{3}{8}$ d. $\frac{1}{2}$ and $\frac{5}{9}$

Unit fractions have a numerator of 1 e.g., $\frac{1}{3}$ and $\frac{1}{5}$. The size of a unit fraction depends on its denominator; **larger** fractions have **smaller** denominators. That is, $\frac{1}{3}$ is larger than $\frac{1}{5}$ since it only requires three thirds (i.e., $3 \times \frac{1}{3}$) to make a whole compared to five fifths (i.e., $5 \times \frac{1}{5}$).

Simplifying fractions

To simplify a fraction, we **rename** it using the smallest possible whole numbers. We do this by dividing the numerator and denominator by the *same* non-zero whole number.

Here we see $\frac{10}{15}$ can be simplified to $\frac{2}{3}$ by dividing both the numerator and denominator by 5.

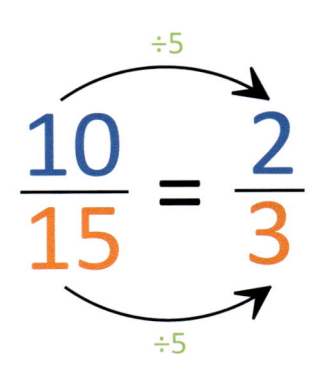

A fraction can be simplified by dividing the numerator and denominator by their highest common factor (HCF) (see glossary). A fraction is in simplest form when the numerator and denominator have no common factors (other than 1).

Exercise 11.1

Simplify the following fractions:

a. $\dfrac{3}{15}$ b. $\dfrac{28}{60}$

c. $\dfrac{16}{28}$ d. $\dfrac{14}{26}$

e. $\dfrac{12}{20}$ f. $\dfrac{8}{36}$

Adding and subtracting like fractions

From their experience with whole numbers, students often expect adding and subtracting fractions to be easier than multiplying and dividing them. It is true that adding and subtracting like fractions is very simple. In these cases, we are simply adding or subtracting a number of equal-sized parts. To add like fractions, simply add the numerators. Do not add the denominators. For example:

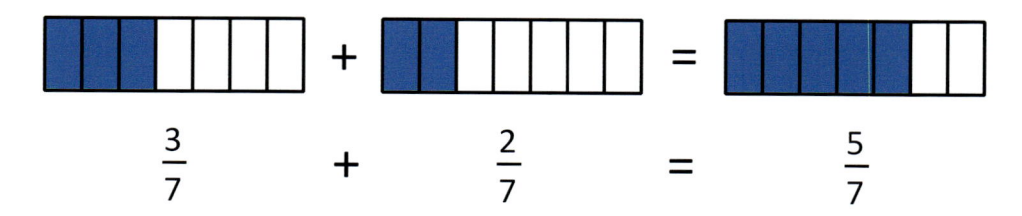

$$\frac{3}{7} \quad + \quad \frac{2}{7} \quad = \quad \frac{5}{7}$$

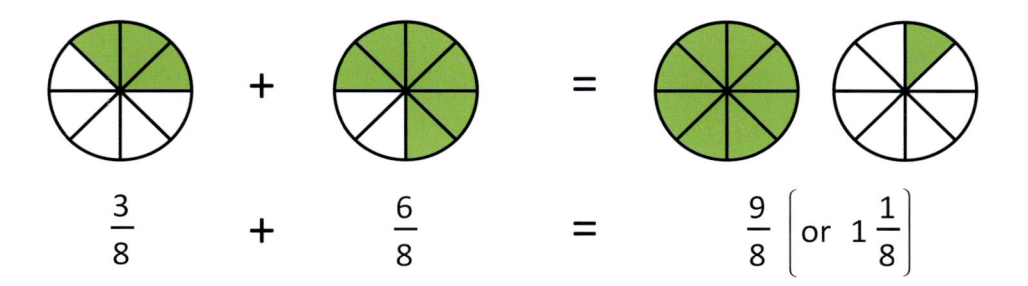

$$\frac{3}{8} \quad + \quad \frac{6}{8} \quad = \quad \frac{9}{8} \left(\text{or } 1\frac{1}{8}\right)$$

To subtract like fractions, we subtract the numerator of the second fraction from the numerator of the first. Do **not** subtract the denominators. For example:

$$\frac{5}{6} \quad - \quad \frac{2}{6} \quad = \quad \frac{3}{6} \left(\text{or } \frac{1}{2}\right)$$

$$\frac{3}{4} \quad - \quad \frac{1}{4} \quad = \quad \frac{2}{4} \left(\text{or } \frac{1}{2}\right)$$

Adding and subtracting unlike fractions

The first step in adding or subtracting **unlike fractions** is to rename them as like fractions, meaning that once again we are working with equal-sized parts. To add unlike fractions, we must first rename the fractions with a common denominator, before adding the numerators. A common denominator is a common multiple of both the original denominators. A simple way to find a common denominator is to multiply the two original denominators together. (Note that this approach does not always give the *lowest* common denominator.)

Example

Calculate $\frac{1}{2} + \frac{1}{3}$

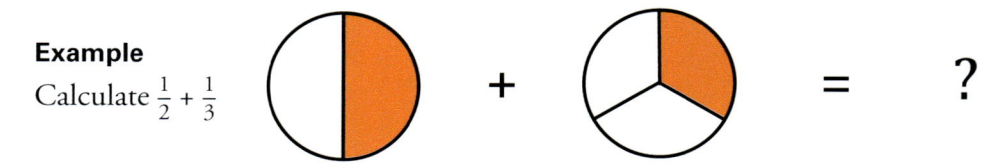

We must first rename the fractions using the common denominator (in this case, 6):

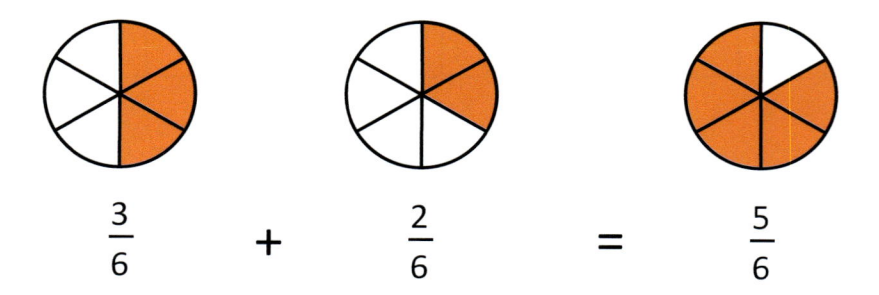

$$\frac{3}{6} \quad + \quad \frac{2}{6} \quad = \quad \frac{5}{6}$$

Since $\frac{1}{2}$ is equivalent to $\frac{3}{6}$ and $\frac{1}{3}$ is equivalent to $\frac{2}{6}$, we can see that $\frac{1}{2} + \frac{1}{3} = \frac{5}{6}$. Note that we do not add the denominators together.

To subtract unlike fractions, we must first rename the fractions with a common denominator, before performing the subtraction. We then subtract the numerator of the second fraction from the numerator of the first. We do not subtract the denominators. For example:

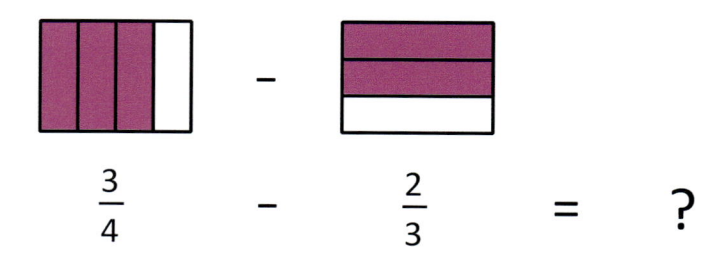

$$\frac{3}{4} \quad - \quad \frac{2}{3} \quad = \quad ?$$

The original fractions are renamed using the common denominator (in this case, 12):

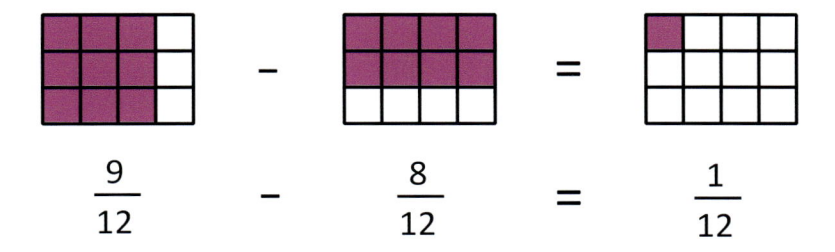

$$\frac{9}{12} \quad - \quad \frac{8}{12} \quad = \quad \frac{1}{12}$$

Since $\frac{3}{4}$ is equivalent to $\frac{9}{12}$ and $\frac{2}{3}$ is equivalent to $\frac{8}{12}$, we can see that $\frac{3}{4} - \frac{2}{3} = \frac{1}{12}$.

Australian Curriculum
Adding and subtracting unlike fractions is introduced in Year 7 of the Australian Curriculum.

Lowest common multiple

Finding the lowest common multiple (or LCM) is useful since it allows fraction calculations to be done using the smallest possible numbers. Using the LCM as the common denominator can greatly simplify the arithmetic required.

Example
Add $\frac{1}{4}$ and $\frac{5}{6}$

Solution

The common denominator is 24 (since $6 \times 4 = 24$). $\qquad \frac{1}{4} + \frac{5}{6}$

We use this to rename both fractions.

$$= \frac{6}{24} + \frac{20}{24}$$

We then add the numerators and simplify.

$$= \frac{26}{24}$$

$$= \frac{13}{12}$$

The final step is to rewrite the improper fraction into a mixed number.

$$= 1\frac{1}{12}$$

Alternatively, if we use the lowest common multiple of 4 and 6 (i.e., 12), we can work with smaller numbers, which leads more directly to the answer:

$$\frac{1}{4} + \frac{5}{6}$$

$$= \frac{3}{12} + \frac{10}{12}$$

$$= \frac{13}{12}$$

$$= 1\frac{1}{12}$$

Exercise 11.2

Complete the following fraction calculations, where possible simplifying your answers.

1. Add the following like fractions:

   a. $\frac{2}{9}$ and $\frac{4}{9}$

   b. $\frac{1}{6}$ and $\frac{4}{6}$

   c. $\frac{3}{12}$ and $\frac{7}{12}$

   d. $\frac{12}{15}$ and $\frac{13}{15}$

2. Subtract the following like fractions:

   a. $\frac{6}{9}$ and $\frac{3}{9}$

   b. $\frac{4}{7}$ and $\frac{2}{7}$

   c. $\frac{13}{12}$ and $\frac{7}{12}$

   d. $\frac{12}{15}$ and $\frac{3}{15}$

3. Add the following unlike fractions:

   a. $\frac{2}{9}$ and $\frac{3}{5}$

   b. $\frac{1}{4}$ and $\frac{3}{7}$

   c. $\frac{5}{12}$ and $\frac{2}{3}$

   d. $\frac{2}{5}$ and $\frac{5}{6}$

4. Subtract the following like fractions:

   a. $\frac{8}{9}$ and $\frac{1}{3}$

   b. $\frac{11}{12}$ and $\frac{5}{6}$

   c. $\frac{5}{8}$ and $\frac{5}{12}$

   d. $\frac{7}{10}$ and $\frac{7}{15}$

Multiplying and dividing fractions

Australian Curriculum
Multiplying and dividing fractions
and mixed numbers is introduced in
Year 7 of the Australian Curriculum.

The procedure for multiplying fractions is simpler than the procedures for adding and subtracting unlike fractions. When multiplying fractions, it is not necessary to find a common denominator. In order to multiply two fractions, we multiply the two numerators to produce the new numerator and multiply the two denominators to produce the new denominator.

Example

What is $\frac{2}{3}$ of $\frac{4}{5}$?

Solution

$$\frac{2}{3} \times \frac{4}{5} = \frac{2 \times 4}{3 \times 5} = \frac{8}{15}$$

Multiplying Mixed Numbers

Mixed numbers must first be converted to improper fractions before they can be multiplied. After the numerator and denominator have been multiplied, the answer can once again be written as a mixed number as shown below.

Example

Find $1\frac{1}{2} \times 2\frac{4}{5}$

Solution

$$1\frac{1}{2} \times 2\frac{4}{5} = \frac{3}{2} \times \frac{14}{5}$$
$$= \frac{42}{10}$$
$$= 4\frac{1}{5}$$

Dividing Fractions

To divide by a fraction, we first rewrite the calculation as a multiplication problem. In other words, to divide by a fraction we turn that fraction upside down and multiply instead. We do this because multiplying fractions is a much simpler process.

Example

Imagine cutting four oranges into eighths for snacks at a football game. The total number of orange slices can be found using one of these methods:

$$4 \div \frac{1}{8} = 32$$

$$\text{or } 4 \times \frac{8}{1} = 32.$$

Exercise 11.3

1. Multiply the following fractions, simplifying where possible:

   a. $\frac{2}{3} \times \frac{4}{5}$

   b. $\frac{3}{7} \times \frac{5}{6}$

   c. $1\frac{3}{5} \times 2\frac{1}{4}$

   d. $1\frac{1}{2} \times \frac{2}{3}$

2. Divide the following fractions, simplifying where possible:

   a. $\frac{2}{3} \div \frac{1}{4}$

   b. $\frac{4}{7} \div \frac{2}{7}$

   c. $\frac{13}{12} \div \frac{1}{6}$

   d. $1\frac{2}{3} \div \frac{1}{9}$

Chapter 12

Decimals and Percentages

Decimal fractions (or simply decimals) involve dividing the whole into ten equal-sized parts (or tenths). Decimals allow the Base 10 system to be extended to include numbers that are between the whole numbers. Each tenth can be subdivided to create hundredths, which can be subdivided to create thousandths and so on. Since decimal fractions are a particular type of fraction, some fraction concepts (such as equivalence) can be extended to decimals. The use of decimal notation means that other fraction concepts (such as common denominators) are entirely unrelated to decimals.

Deci is a prefix meaning one-tenth. The word 'decimate' originally meant killing one in ten members of a Roman legion as a means of punishing the whole group.

Decimal notation

Representing numbers using decimal notation is highly efficient. We use the same Base 10 place value system as we do for whole numbers but extend it to include digits representing the fractional parts of the number. A decimal point is used to separate the whole numbers from the fractional part of the number. For this reason, teachers should avoid referring to 'moving the decimal point' since it is always positioned between the whole number part and the fractional part.

Whole Number Part				Fractional Part		
Hundreds	Tens	Ones		Tenths	Hundredths	Thousandths
3	8	1	.	0	2	7

Notice that the decimal place values (i.e., tenths, hundredths and thousandths) do not mirror the whole number place values (i.e., ones, tens and hundreds).

Comparing decimals

Just as with whole numbers, place value is used to compare decimals. To decide whether one number is larger than another, we look at the left hand digit of both numbers (i.e., the digit with the largest place value). If both digits have the same value, we move to the next largest place value column and continue until we find one digit which has a larger value than the other.

In the example below, we are comparing 12.191, 12.19 and 12.094. First, look at the left hand digits (the tens column. Since the digits in the tens column are the same, we move to the next largest place value column (i.e., ones). Since these are also the same, we must compare the fractional parts of the numbers. By comparing the digits in the tenths column, we see that both 12.191 and 12.19 are greater than 12.094, since they each have more tenths.

To compare 12.191 and 12.19 we must look at the thousandths column. We see that 12.191 is larger than 12.19 because it possesses more thousandths. Note that we can also write 12.19 as 12.190 without altering its value. Sometimes this is a useful strategy when comparing decimals.

Tens	Ones		Tenths	Hundredths	Thousandths
1	2	.	1	9	1
1	2	.	1	9	
1	2	.	0	9	4

Using expanded notation, these numbers can be expressed as:

$$12.191 = 10 + 2 + \frac{1}{10} + \frac{9}{100} + \frac{1}{1000}$$

$$12.19 = 10 + 2 + \frac{1}{10} + \frac{9}{100} + \frac{0}{1000}$$

$$12.094 = 10 + 2 + \frac{0}{10} + \frac{9}{100} + \frac{4}{1000}$$

Rounding decimals

Australian Curriculum
Rounding decimals to a specified number of places is introduced in Year 7 of the Australian Curriculum; however, rounding of whole numbers and decimals more generally is introduced in the primary years.

The purpose of rounding is to make numbers easier to handle by reducing the number of decimal places while retaining the same approximate value. When rounding a number, we must first decide on the required accuracy (for example, two decimal places). The number in the next decimal place (in this case, the third decimal place) then determines whether we round up to the next number or down to the preceding number.

Example 1

Round 2.71828 to two decimal places.

Solution

In order to round to the required accuracy, we look at the digit in the third decimal place (in this case, 8).

$$2.71 | 828$$

If the digit is 5 or greater, we round up to the next hundredth, which is 2.72.

Example 2

Round 6.1449 to two decimal places.

Solution

In order to round to the required accuracy, we look at the digit in the third decimal place (in this case, 4). Since 4 is less than 5, we round down to the nearest hundredth, which is 6.14.

$$6.14 | 49$$

Note: It is incorrect to round repeatedly. For example, do not first round to three decimal places (i.e., 6.145) before rounding to two decimal places (i.e., 6.15).

Example 3

Round 214.017 to the nearest hundredth.

Solution

To the nearest hundredth, 214.017 will be either 214.01 or 214.02. The number line representation makes it clear that 214.017 is closer to the 214.02 and so in this case we round up.

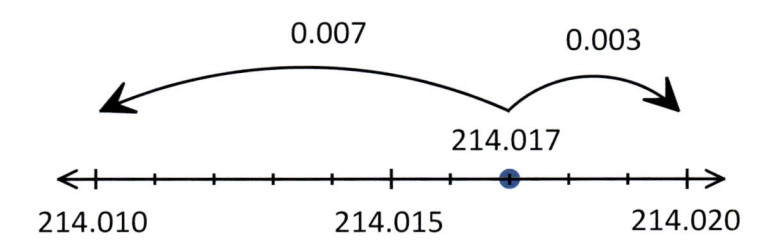

Exercise 12.1

Round the following decimals to the required accuracy.

a. 12.87 to the nearest tenth

b. 216.0079 to the nearest hundredth

c. 0.0186 to the nearest thousandth

d. 49.087 to one decimal place

e. 0.02479 to three decimal places

f. 999.987 to two decimal places

Adding decimals

The procedure for adding decimals is an extension of the procedure for adding whole numbers. Once again, two (or more) numbers are aligned according to their place values. The decimal point in the answer is aligned with the decimal points in the numbers being added. Calculation proceeds from right to left, adding the smallest decimal component first. We add the digits and record the total. If the total exceeds 9, then we record the ones digit and carry 1 (or more) across to the next column on the left. We then proceed to add the digits in the next column to the left. Again we record the total and if it exceeds 9, we record the ones digit and carry 1 (or more) across to the next column on the left. We continue this process until all of the digits have been added.

Example

In order to add 3.26 to 1.81, we first align the place value columns. We begin from the right, adding 6 hundredths and 1 hundredth to get 7 in the hundredths column. Moving to the left, we recognise that 2 tenths and 8 tenths creates 10 tenths, which is exchanged for one whole. We record a zero in the tenths column and add the additional whole to the ones column, giving a total of 5.07 as shown below.

$$
\begin{array}{r}
{}^{1}3 \cdot 2\ 6 \\
+\ 1 \cdot 8\ 1 \\
\hline
5 \cdot 0\ 7 \\
\hline
\end{array}
$$

Exercise 12.2

Work through the following examples of the written addition algorithm. No calculators!

a.
```
    1 · 6
  + 3 · 9
  ───────
      ·
```
b.
```
    2 · 7 4
  + 3 · 1 6
  ─────────
      ·
```
c.
```
      9 · 6 7
  + 1 0 · 8 9
  ───────────
        ·
```

d.
```
      1 · 2
      3 · 4
    1 2 · 7
  + 2 0 · 9
  ─────────
        ·
```
e.
```
    0 · 4
    2 · 9
    1 · 6
  + 3 · 5
  ───────
      ·
```
f.
```
    0 · 0 8 7
    0 · 4 1 9
    1 · 2 1 6
  + 2 · 4 9 1
  ───────────
        ·
```

Subtracting decimals

The procedure for subtracting decimals is also an extension of the procedure for subtracting whole numbers. Once again, we align the two numbers according to their place values, with the larger number on top. The decimal point in the answer is aligned with the decimal points in the numbers being added. Beginning with the right hand column, we subtract the bottom number from the top number. (This will require exchanging or regrouping if the bottom number is smaller than the top number.) Record the single-digit answer and move to the next column to the left.

Example

In order to subtract 2.64 from 4.09, we first align the place value columns. We begin from the right, subtracting 4 hundredths from 9 hundredths to get 5 hundredths. Moving to the left, we recognise that we must exchange one whole for 10 tenths, since 6 tenths cannot be subtracted from zero tenths. Subtracting 6 tenths from 10 tenths leaves 4 tenths. Finally, we subtract 2 ones from the remaining 3 ones to get 1 in the ones column.

$$
\begin{array}{r}
^3\cancel{4} \cdot {}^1 0 \ \ 9 \\
- \ \ 2 \cdot \ \ 6 \ \ 4 \\
\hline
1 \cdot \ \ 4 \ \ 5 \\
\hline
\end{array}
$$

Exercise 12.3

Work through the following examples of the subtraction algorithm.

```
a.      7 · 1      b.    3 · 4 1    c.    2 · 9 3 7
      − 2 · 9          − 2 · 8 6        − 1 · 0 8 9
      ─────────        ─────────        ───────────
           ·                ·                ·
      ─────────        ─────────        ───────────


d.      3 · 4 9    e.    7 · 0 6    f.    2 · 2 1 0
      − 2 · 9 1          − 3 · 4 1        − 1 · 0 9 1
      ─────────        ─────────        ───────────
           ·                ·                ·
      ─────────        ─────────        ───────────
```

Multiplying decimals

 **Australian Curriculum**
Multiplication and division of a decimal by a whole number is introduced in Year 6. Multiplying and dividing a decimal by a decimal is introduced in Year 7 of the Australian Curriculum.

The procedure for multiplying decimals is also similar to the procedure for multiplying whole numbers. The two factors being multiplied are aligned according to their place values, with the decimal points aligned. The number with the most digits is generally placed on top (although this is not a requirement). Calculation proceeds from right to left. Multiply each of the digits of the top number (beginning with the right hand digit and proceeding to the left) by the right hand digit of the bottom number, writing down the ones digit of the product and carrying where necessary.

We then move to the second from right digit of the bottom number. We place a zero as a placeholder in the right hand column underneath our previous calculations. We then multiply each of the digits of the top number (beginning with the right hand digit and proceeding to the left) by the second from right digit of the bottom number, in each case writing down the ones digit of the product and carrying where necessary.

If necessary, we then turn our attention to the third from right digit of the bottom number. We place two zeroes as placeholders in the right hand positions of the row underneath our previous calculations. We then multiply each of the digits of the top number (beginning with the right hand digit and proceeding to the left) by the third from right digit of the bottom number, in each case writing down the ones digit of the product and carrying where necessary.

We continue in this manner until we have calculated all of the partial products. The final step is to add these partial products, taking care to ensure that the place values are aligned correctly, to get the final answer. To place the decimal point in our answer, we add the number of decimal places in the two products. For example, if we multiply 4.32×1.4, we will have a total of three decimal places in our product i.e., 6.048.

Example

Calculation of 4.32×1.4 using the written algorithm:

			${}^{1}4$	·	3	2
		×	1	·	4	
This row shows $4.32 \times 0.4 \rightarrow$		${}^{1}1$	7		2	8
This row shows $4.32 \times 1 \rightarrow$		4	3		2	0
This row shows $4.32 \times 1.4 \rightarrow$		6	·	0	4	8

Exercise 12.4

Work through the following examples of the standard written algorithm for multiplication. No calculators!

a.　　0 · 2 8　b.　　7 · 4　c.　　6 · 1 8
　　x 0 · 3 1　　　x 1 · 6　　　x 0 · 3 4

Dividing decimals

The procedures for the division of whole numbers also apply to decimals. Divisions involving finite decimals can be converted to divisions involving whole numbers by multiplying both the dividend and divisor by a suitable power of ten. For example, $2.4 \div 0.3$ is equivalent to $24 \div 3$. While both divisions have the same answer, whole number division is usually much easier to perform.

Occasionally a division will produce an answer where the decimal digits endlessly repeat. For example, $20 \div 3$.

$$\begin{array}{c} \ \ \ 6 \ \cdot \ 6 \ \ 6 \\ \hline 3 \ | \ 2 \ \ 0 \ \cdot \ 0 \ \ 0 \end{array}$$

This is known as a *recurring decimal* and is shown as $6.\overline{6}$ with a line above the number that repeats. You may also see a dot used above the number rather than a line. (See later section on Rational and Irrational Numbers).

Exercise 12.5

Rewrite the following decimal divisions into problems involving whole numbers by multiplying by a suitable power of 10:

a. 3.8 ÷ 1.6
b. 6.23 ÷ 3.81
c. 1.5 ÷ 0.25
d. 4.23 ÷ 1.41
e. 1.633 ÷ 0.88
f. 0.0016 ÷ 0.008

Percentages

In the same way decimals can be thought of as fractions involving powers of 10, percentages can be thought of as fractions of 100. The term 'per cent' is derived from the Latin *per centum* which means 'by the hundred'. Percentages are useful for making comparisons. For example, percentages are often used to compare student grades, interest rates, tax rates, price increases or discounts, the chances of rain and relative humidity.

Decimals and percentages may be illustrated using a 10 × 10 grid. Each row or column that is shaded represents one-tenth or 10% of the whole. Here we can see that 25% is equivalent to one-quarter of the total.

It is important to realise that percentages are not restricted to values between 0 and 100%. For example, 117% can be represented as shown in the diagram below.

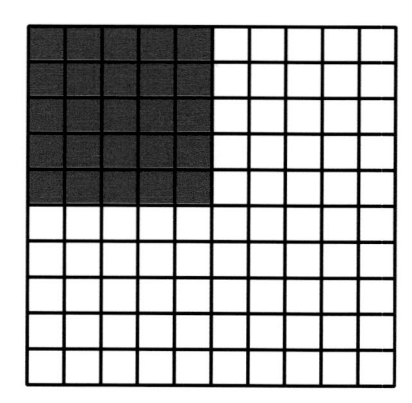

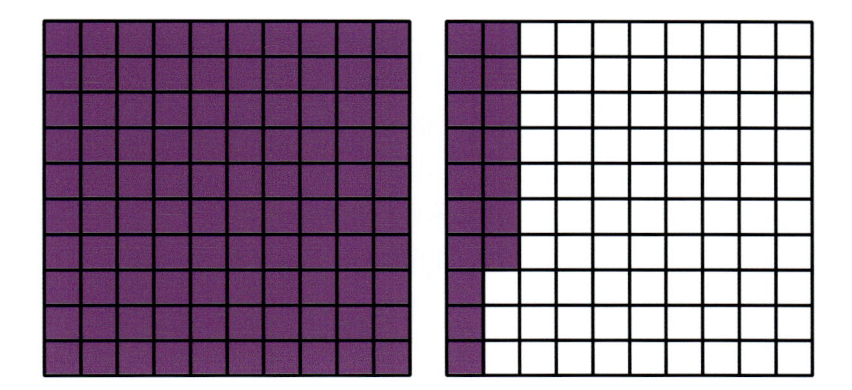

Decimals and percentages can also be represented on a number line. Notice that the number line continues beyond the range of 0–1 (for decimals) and 0–100% for percentages.

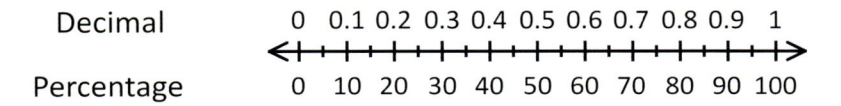

Conversions

There are six ways to convert between fractions, decimals and percentages. Converting between decimals and percentages involves multiplying and dividing by 100, which is easy in the Base 10 system. For example, 0.68 becomes 68% and 1.25 becomes 125%.

To convert a percentage to a fraction, we use the percentage amount as the numerator and place it over a denominator of 100. It is then often necessary to simplify the resulting fraction. For example, 65% can be written as $\frac{65}{100}$ which simplifies to $\frac{13}{20}$. A fraction may be converted into a percentage by expressing it with a denominator of 100. For example, $\frac{4}{5}$ can be renamed as $\frac{80}{100}$ which is equivalent to 80%. In order to do this, we work out how many times 5 goes into 100 (i.e., 20) and multiplying both the numerator and denominator by this amount. (**Note:** not all denominators divide evenly into 100 and so it is sometimes easier to convert a fraction to a decimal before converting it to a percentage.)

We use division to convert fractions into decimals. Generally, this is done using a calculator but it can also be done using long or short division. For example, to convert $\frac{5}{8}$ into a decimal, we divide 5 by 8 (i.e., 5 ÷ 8), which gives 0.625. A context for this might be sharing 5 pizzas equally between 8 people. A quick estimation should tell you that each person will get a bit more than half a pizza each.

Finally, we use decimal place values to convert decimals into fractions. For example, 4.32 can be converted to $4\frac{32}{100}$ using place value and this simplifies to $4\frac{8}{25}$.

The diagram below summarises fraction, decimal and percentage conversions.

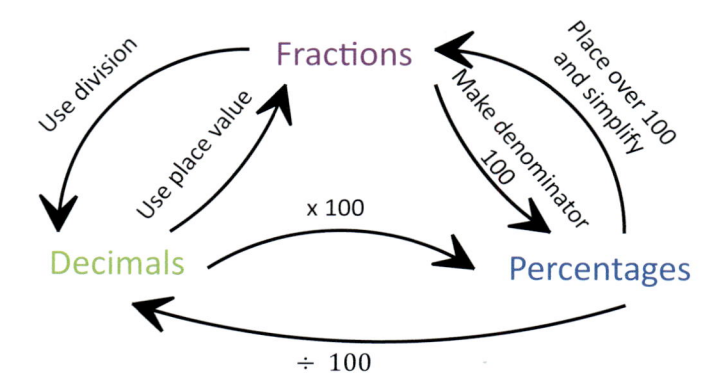

Rational and irrational numbers

Fractions are numbers which can be written as the ratio of two whole numbers. For this reason, numbers which can be expressed as fractions are called rational numbers. Any rational number (including improper fractions) can be converted into a decimal by simply dividing the numerator by the denominator.

Decimal numbers that cannot be written as fractions are known as irrational numbers. In other words, these numbers cannot be expressed as the ratio of two whole numbers. Examples include mathematical constants as π, e and ϕ (the golden ratio). Each of these numbers is an infinite, non-repeating decimal. For example, the value of π is *approximately* 3.14159265 (but not exactly).

Exercise 12.6

1. Convert the following decimals to percentages:
   - a. 0.16
   - b. 0.425
   - c. 1.01
   - d. 0.005
2. Convert the following percentages to decimals:
   - a. 38%
   - b. 141.8%
   - c. 67%
   - d. 1.09%
3. Convert the following decimals to fractions:
   - a. 0.34
   - b. 0.07
   - c. 2.08
   - d. 1.19
4. Convert the following fractions to decimals:
   - a. $\frac{12}{20}$
   - b. $1\frac{1}{5}$
   - c. $\frac{11}{12}$
   - d. $\frac{13}{30}$
5. Convert the following percentages to fractions or mixed numbers:
   - a. 64%
   - b. 5.9%
   - c. 110%
   - d. 12.1%
6. Convert the following fractions to percentages:
   - a. $\frac{9}{20}$
   - b. $\frac{5}{8}$
   - c. $2\frac{1}{4}$
   - d. $\frac{17}{40}$

Chapter 13

Measurement

Measurement is the process by which we quantify observable aspects of the world around us. These aspects, known as *attributes*, include mass, length, time, area, volume/capacity, temperature and angles. The measurement process involves first identifying the attribute to be measured and then selecting a unit against which the attribute can be compared e.g., a metre, a litre or a kilogram (Siemon et al., 2011).

There are many other units of measurement that are not part of the metric system. These include: pounds, gallons, acres, miles, yards, feet and inches. For the purpose of primary classroom teaching, we focus on the metric system.

Measuring instruments

Different instruments are used to measure each attribute. Length is measured using rulers, tape measures, trundle wheels, calipers, micrometers and

Attribute	Dimensions	Standard Units
Length	1	Linear units e.g., centimetres (cm)
Area	2	Square units e.g., square centimetres (cm^2)
Volume	3	Cubic units e.g., cubic centimetres (cm^3)
Capacity	3	Litres (L) or millilitres (mL)
Mass		Kilograms (kg) or grams (g)
Time		Seconds (s), minutes (min), or hours (h)
Temperature		degrees Celsius
Angle		degrees

odometers. Mass can be measured with scales (or compared using a balance). Time is measured using watches, clocks, calendars, digital timers and sundials. Temperature is measured using thermometers and angles are measured using protractors. Other attributes, such as area and volume, are often derived from measures of length, width and height.

A fundamental principle of measurement is that attributes can never be determined with complete accuracy. In other words, when we measure an attribute such as length, the value obtained is only approximate. For example, if we measure a table using a metre ruler with centimetre divisions, we can only claim that the measurement is accurate to the nearest centimetre. If we were to measure the same table using a tape measure with millimetre divisions, we can still only claim accuracy to the nearest millimetre and so on.

The required level of accuracy of a particular measurement often depends on the purpose for its use. For example, it would be reasonable to express the length of a car journey to the nearest kilometre, but not to the nearest centimetre or millimetre. The level of accuracy required can also dictate the type of measuring instrument that is used. For example, an engineering toler-ance might need to be measured to the nearest hundredth of a millimetre and so using a ruler or tape measure would be inappropriate.

Activity

Children need to learn to use measuring instruments accurately and this involves being able to read and interpret the graduations on scaled instruments.

1. Have a look around your home for items that have a measuring scale such as rulers, tape measures, measuring jugs, buckets, droppers, oven temperature, thermometers, etc. Examine the scale on each. What unit of measurement is used? What do the intervals between each unit represent (i.e., the lines between each unit that are often unlabelled)?

2. What mathematical skill did you use to calculate what each interval represented?

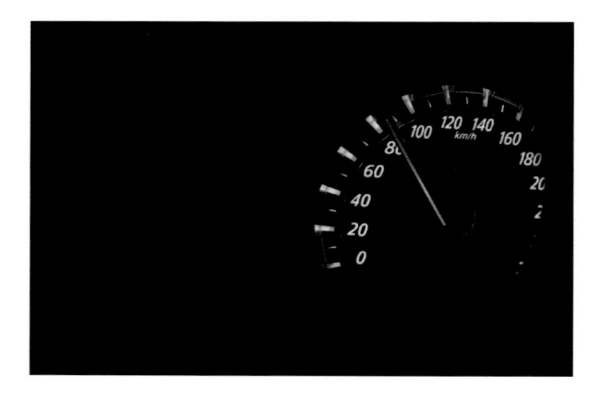

Length

With the possible exception of time, length is the attribute that we are most familiar with measuring. Probably the first measurement tool children encounter is the ruler. Using a ruler is more difficult than one might think. In order to measure the length of a table, we need to understand that:

- The ruler must be aligned to the edge of the table.
- The length is measured directly across the table (not diagonally).
- There can be no gaps or overlap between the units.

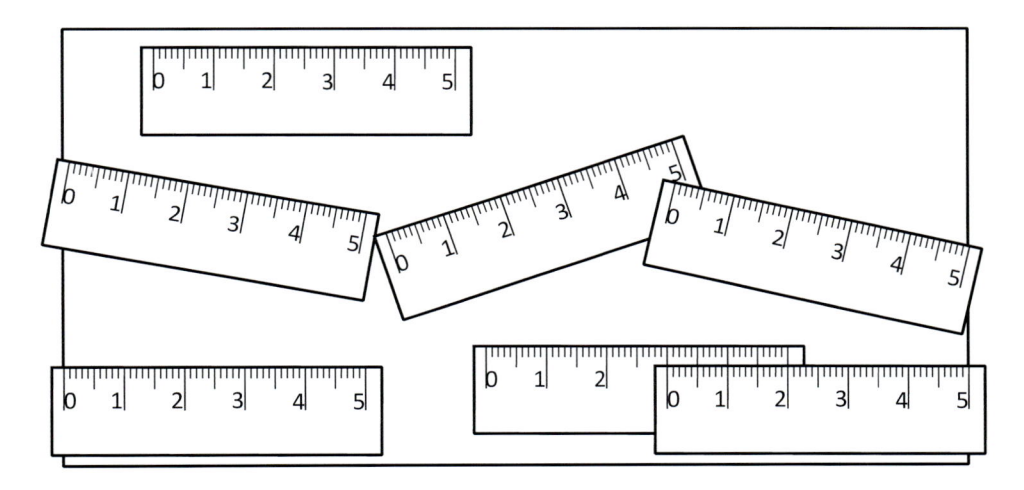

In the metric system, length is measured using kilometres, metres, centimetres and millimetres. A kilometre is 1 000 metres. A metre consists of 100 centimetres and 1 000 millimetres.

Area

Area quantifies the size of a two-dimensional surface. We use area to work out the amount of a wall surface to be painted or the amount of cardboard required to construct a cereal box. Area is generally found by multiplying together two measures of length e.g., the length and width or the base and height.

Area is measured in square kilometres, square metres, square centimetres and square millimetres. A square kilometre has an area equivalent to 1 000 000 square metres, since it is a square 1 000 metres long by 1 000 wide. A square metre has an area equivalent to 10 000 square centimetres, since it is a square 100 centimetres by 100 centimetres. A square centimetre has an area equivalent to 100 square millimetres, since it is a square 10 millimetres by 10 millimetres.

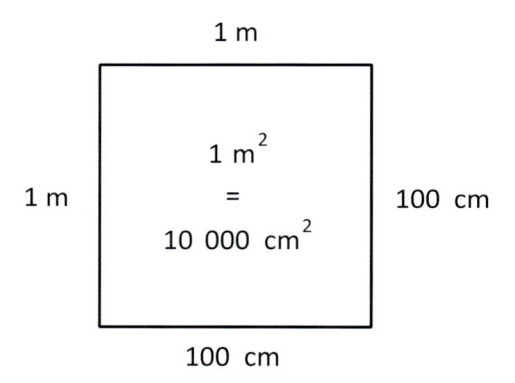

Volume and capacity

Volume refers to the amount of space taken up by a three-dimensional object, while capacity is the amount that a container can hold. Volume is generally found by multiplying together three measures of length i.e., length, width and height. It can also be found indirectly by measuring the amount of water an object displaces when it is submerged (i.e., Archimedes' principle). The capacity of a container is usually determined by measuring how much liquid it can hold.

Volume is measured in cubic kilometres, cubic metres and cubic centimetres. A cubic kilometre has a volume equivalent to 1 000 000 000 cubic metres, since it is a cube 1 000 metres by 1 000 metres by 1 000 metres. A cubic metre has a volume equivalent to 1 000 000 cubic centimetres, since it is a cube 100 centimetres by 100 centimetres by 100 centimetres. A cubic metre has a capacity of 1 000 litres, while a cubic centimetre has a capacity of 1 millilitre (mL).

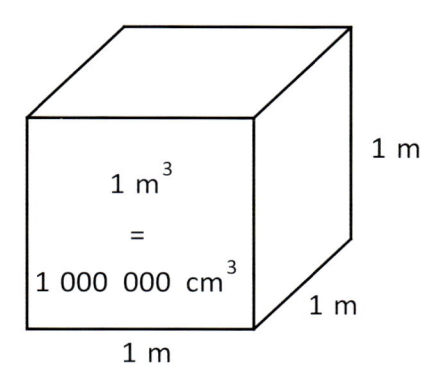

Mass

Mass refers to the amount of matter that is contained in a solid object. Mass is often confused with weight, which is the force which gravity exerts on the object. Mass is measured in tonnes, kilograms, grams and milligrams. There are 1 000 kilograms in a tonne. There are 1 000 grams in a kilogram and 1 000 milligrams in a gram.

Time

Time is measured using a wide variety of devices and expressed in a multitude of units. For example, we refer to seconds, minutes, hours, days, weeks, fortnights, months, seasons, years, decades, centuries and millennia. We use time to refer not only to when a particular event happens but also the duration of the event.

Temperature

It is very common for people to refer to temperature in a subjective manner, such as when they are feeling hot or cold. A more precise definition is required in order to measure temperature in an objective manner. Scientifically speaking, temperature is a measure of the average amount of heat possessed by the particles in a sample.

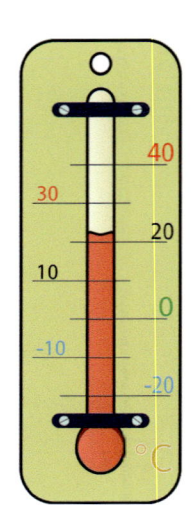

Temperature is measured in degrees Celsius. The difference between the freezing point of water (i.e., 0° C) and the boiling point of water (i.e., 100° C) is divided into one hundred equal intervals, each of which is one degree.

Angles

Angles are measures of the amount of *turn* and are measured in degrees (i.e., °). A complete turn (or revolution) is divided into 360°. Angles are usually classified by comparing their size in relation to a *right angle*. A right angle is a quarter turn and measures 90°.

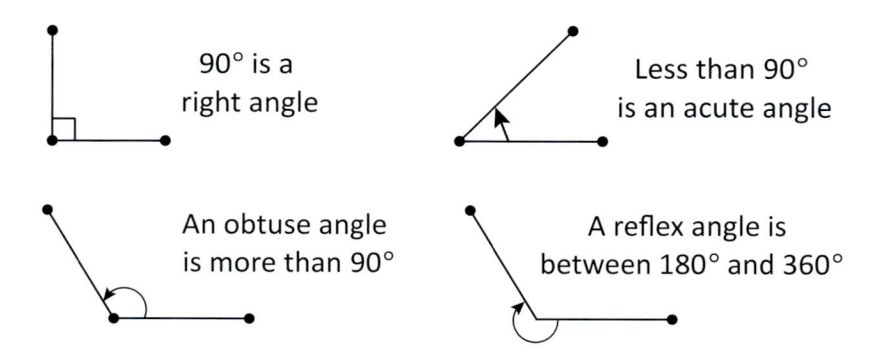

To measure an angle, we place the centre of the protractor on the *vertex* of the angle (i.e., the point where the two *rays* meet), ensuring that the baseline of the protractor is aligned with the lower of the two rays. In the example below, the angle shown is 45°. Since we are measuring the amount of turn in an anti-clockwise manner, we use the inner scale which runs from right to left.

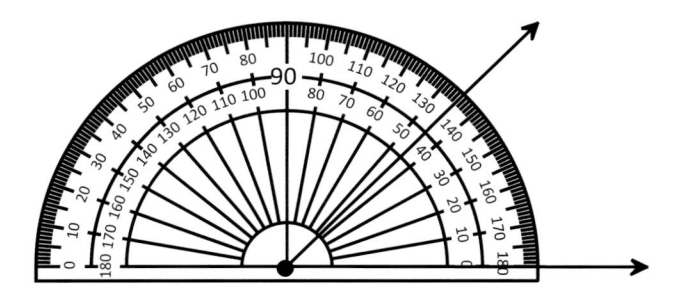

In the following example, the angle shown is 114°. Since we are measuring the amount of turn in a clockwise manner, we use the outer scale which runs from left to right.

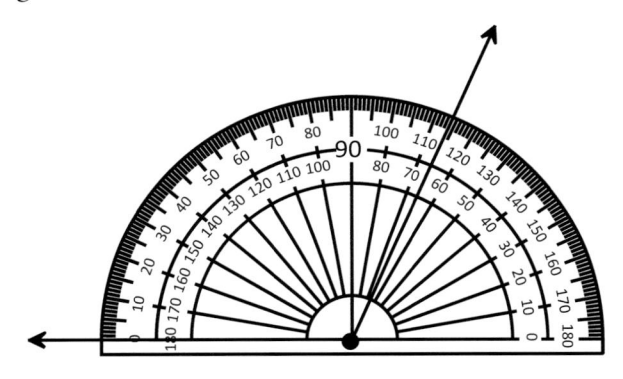

Exercise 13.1

Use a protractor to measure the following angles:

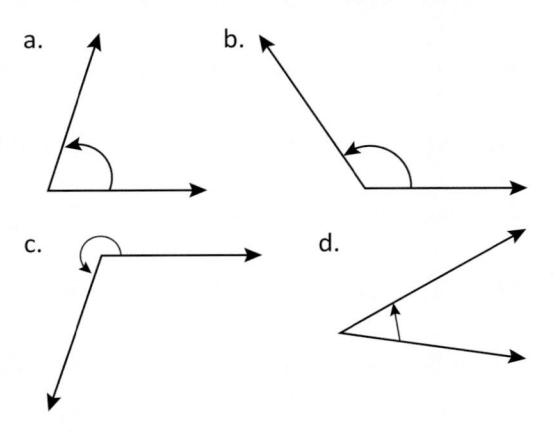

Conservation and conversion

A complete measurement consists of a number of units that match the attribute that is being measured. For convenience, it is sometimes preferable to express measurements using larger or smaller units. This process is known as conversion. When a measurement is converted from one unit to another, the measurement is conserved. For example, if the length of a table is 1 200 mm, the length can also be expressed as 120 cm or 1.2 metres. The length of the table does not change, since 1 200 mm is equivalent to 120 cm and 1.2 m.

When we convert from a larger unit (e.g., kilometres) to a smaller unit (e.g., metres), we multiply. Since there are 1 000 metres in a kilometre, we need to multiply by 1 000 in order to convert kilometres to metres. When we convert from a smaller unit (e.g., grams) to a larger unit (e.g., kilograms), we divide. Since there are 1 000 grams in a kilogram, we need to divide by 1 000 in order to convert grams to kilograms.

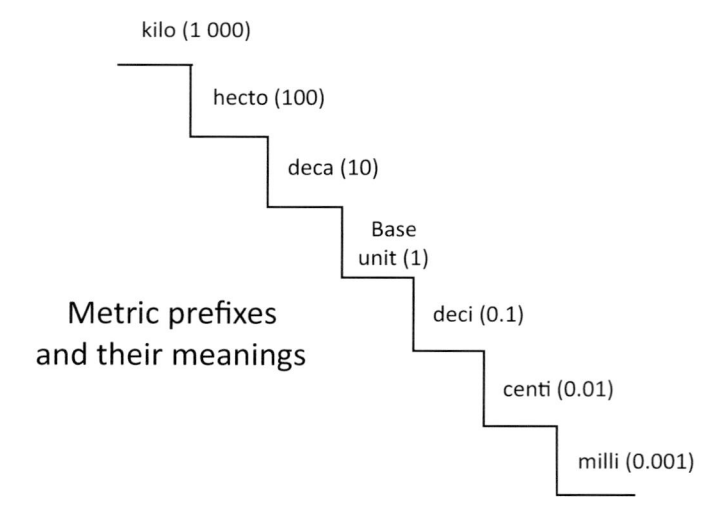

In the conversion staircase shown, larger units are positioned above smaller units. Conversions can be performed by moving up and down the staircase. Each time we take a step up, we must divide our measurement by 10. For example, when converting from milli-units to centi-units, we need to move up one step and so must divide by 10. This would enable us to convert millimetres to centimetres.

In order to convert larger units to smaller units, we must move down steps on the staircase. Each time we move down a step, we must multiply by 10. For example, if we were to convert kilo-units to base-units, we would need to multiply by 10 *three times* (i.e., $10 \times 10 \times 10 = 1\ 000$). This would enable us to convert kilometres to metres, or kilograms to grams, or kilolitres to litres.

Exercise 13.2

1. Perform the following length conversions:
   a. 12.5 m to cm
   b. 1 117 mm to cm
   c. 3.4 km to m
   d. 0.005 km to cm
2. Perform the following mass conversions:
   a. 12.05 kg to g
   b. 1 355 kg to t
   c. 468 g to kg
   d. 25 000 mg to g
3. Perform the following volume conversions:
   a. 120 L to kL
   b. 3 450 mL to L
   c. 0.06 kL to L
   d. 45.5 L to mL
4. Perform the following time conversions:
   a. 4 fortnights to days
   b. 3 days to hours
   c. 7.5 hours to minutes
   d. 1 year to hours
5. Challenge. Perform the following area conversions:
   a. 100 m^2 to km^2
   b. 1 000 mm^2 to cm^2
   c. 25 000 cm^2 to m^2
   d. 0.005 km^2 to m^2

Chapter 14

Geometry

Geometry is the study of location, space and shapes in both two and three dimensions. Typically we refer to the branch of geometry that deals with two-dimensional *figures* and *shapes* as plane geometry, while the study of three--dimensional *objects* is called solid geometry. Studying geometry allows us to find perimeters, areas, volumes and angles, and to better understand the world in which we live.

Polygons

Polygons are simple closed plane figures with three or more straight sides. **Regular** polygons are those which have all sides and angles the same size. For example, a square is a regular polygon but a right-angled triangle is not.

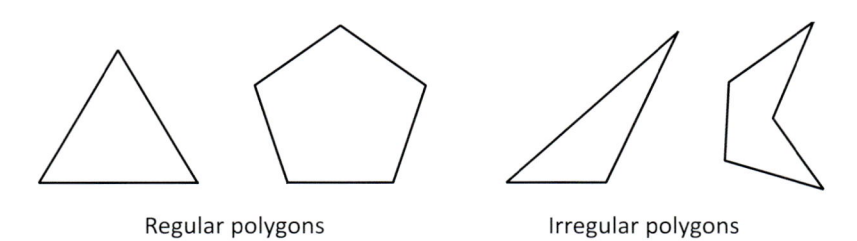

Regular polygons   Irregular polygons

Polygons with three sides and three angles are called triangles, while those with four sides are called quadrilaterals. Triangles and quadrilaterals are classified according to their side and angle properties. Polygons with more than four sides are named by the number of sides they have. Pentagons have five sides, hexagons have six, septagons have seven, octagons have eight and so on.

Although there are mathematical names for polygons with large numbers of sides, it is acceptable to call them *n*-gons when *n* is sufficiently large. For example, a 42-sided polygon is called a 42-gon! It may seem inconsistent that we refer to *triangles* rather than *trigons*, but we do use *trigonometry* to refer to the branch of mathematics that concerns measuring with triangles. While you may have spent some time studying *quadrilaterals* in mathematics, you are likely to have spent a great deal more time in the school's *quadrangle*.

Triangles

Triangles which have all sides the same length and all angles the same size are called equilateral. In the diagram on the right, tick marks have been used to indicate sides of the same length. Since the three angles in any triangle must add to 180°, each angle in an equilateral triangle is 60°.

Isosceles triangles have two sides the same length (shown using the // symbol on the line). The angles opposite these two sides are the same size.

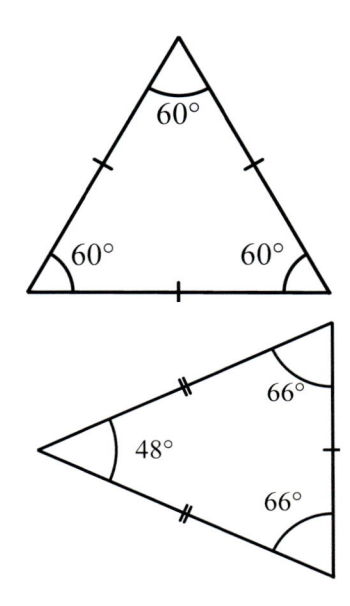

Scalene triangles have no sides the same length and no angles the same size. Each line has been marked with a different symbol.

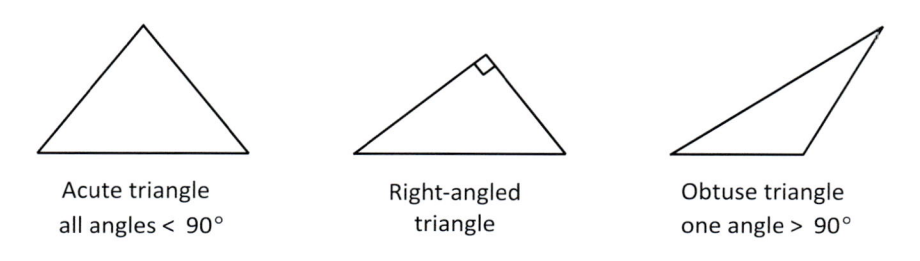

Triangles can also be classified by their angle sizes. All of the angles in acute triangles are between 0 and 90°, right-angled triangles have one 90° angle and obtuse triangles have one angle that is between 90 and 180°.

| Acute triangle | Right-angled | Obtuse triangle |
| all angles < 90° | triangle | one angle > 90° |

Activity

Is it possible to classify triangles by both their side and angle properties? For example, is there such a thing as an acute isosceles triangle? If so, what other combinations are possible?

Quadrilaterals

Quadrilaterals are classified by their side and angle properties. Parallelograms are the broadest class of quadrilaterals and include rhombuses, rectangles and squares. All parallelograms have two pairs of parallel sides, with rectangles also containing right angles, while squares and rhombuses have all sides of equal length. In contrast, trapeziums have just one pair of parallel sides and kites have no parallel sides.

The table below illustrates the different classes of quadrilateral and their properties. Notice the way symbols have been used to indicate sides that are

Quadrilateral	Properties	Diagram
Parallelogram	• Opposite sides parallel • Opposite sides equal	
Rectangle	• Opposite sides parallel • Opposite sides equal • Contains four right angles	
Square	• Opposite sides parallel • All sides equal • Contains four right angles	
Rhombus	• Opposite sides parallel • All sides equal	
Trapezium	• One pair of parallel sides	
Kite	• Two pairs of equal sides that are adjacent (i.e., next to)	

the same length and angles that are the same size. In the trapezium, arrows have been used to show that one pair of sides are parallel.

The angles in any quadrilateral add up to 360°. Notice that any quadrilateral can be divided into two triangles by drawing a line joining two non-adjacent vertices (i.e., a **diagonal**). The two triangles formed in each case are not necessarily the same size.

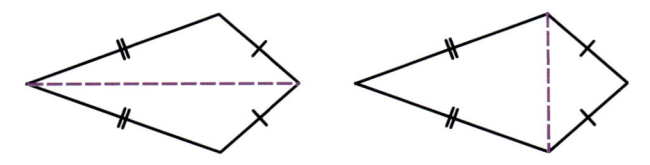

Perimeter

Perimeter is the distance around the boundary of a two-dimensional shape. In the case of polygons it is found by adding together the lengths of the individual sides. For regular polygons (such as squares) the perimeter is found by multiplying the side length by the number of sides. The perimeter of a rectangle and regular hexagon are calculated in the example below.

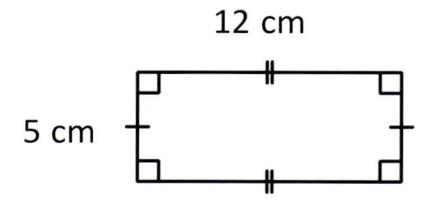

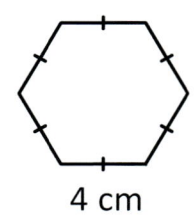

Perimeter = 2 × 12 + 2 × 5

= 34 *cm*

Perimeter = 6 × 4

= 24 *cm*

Australian Curriculum
Calculating the perimeter and area of rectangles is introduced in Year 5 but more complex shapes such as circles, parallelograms, trapeziums, rhombuses and kites are not introduced until Years 7 and 8.

The perimeter of a circle is called the **circumference**. The circumference of a circle is proportional to its **diameter**:

$$C = \pi \times d$$

The ratio of the circumference of a circle to its diameter is called **pi** (π). The value of pi is approximately 3.14 (or $3\frac{1}{7}$). Remember pi

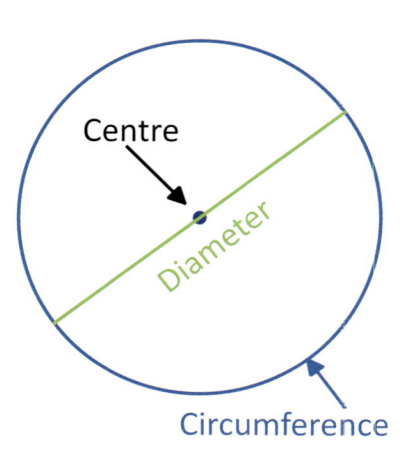

is an irrational number so it does not have an exact decimal representation.

A circle with a diameter of 10 cm therefore has a circumference of $10 \times \pi \approx 31.4 \, cm.$

Note: The perimeter of many shapes can be found using a combination of polygons (e.g., triangles and rectangles) and circles.

Exercise 14.1

Calculate the perimeter of the following shapes. Where necessary, give your answer correct to two decimal places. (Use $\pi = 3.14$ where necessary.)

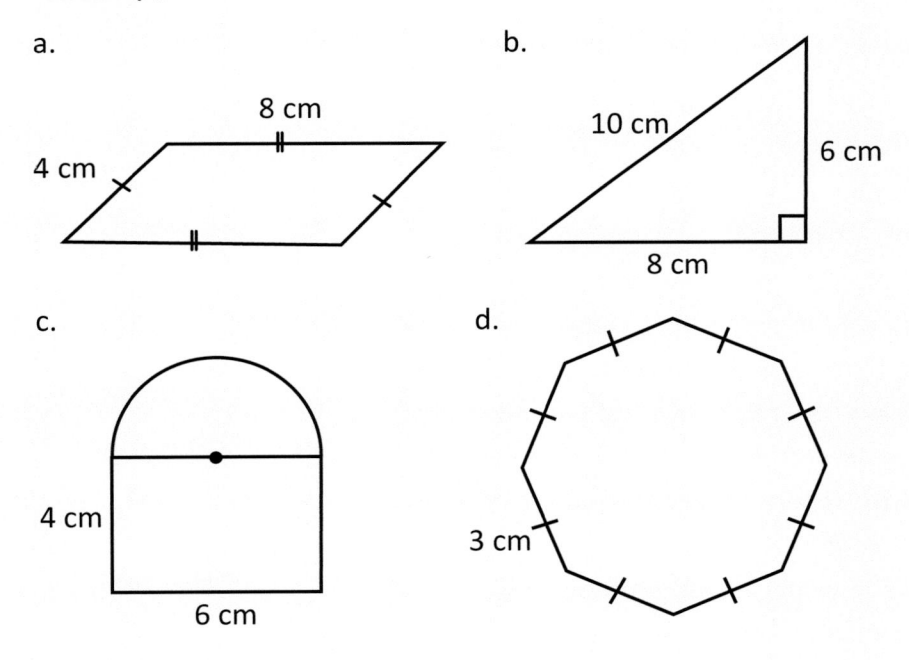

a.

8 cm

4 cm

b.

10 cm

6 cm

8 cm

c.

4 cm

6 cm

d.

3 cm

Area

Area is the size of a two-dimensional surface. It may help to think of the area of a surface as being the amount of paint that would be needed to cover the entire surface. To calculate the area of a surface we need two measurements—either the length and width (in the case of rectangles) or the base and height (in the case of triangles and parallelograms). In the example below, the area of a rectangle and a parallelogram have been calculated.

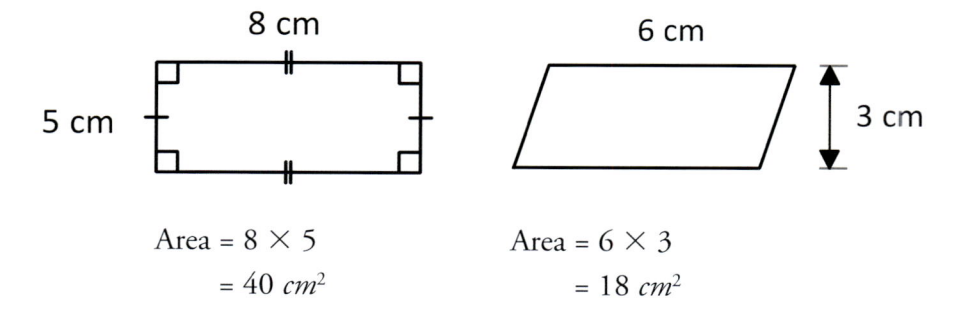

Area = 8 × 5
= 40 *cm²*

Area = 6 × 3
= 18 *cm²*

Since area involves measuring two dimensions, it is expressed using square units e.g., square metres (m²), square centimetres (cm²) or square kilometres (km²). When calculating area, the two measurements must be at right angles to one another (this is why we sometimes refer to the perpendicular height).

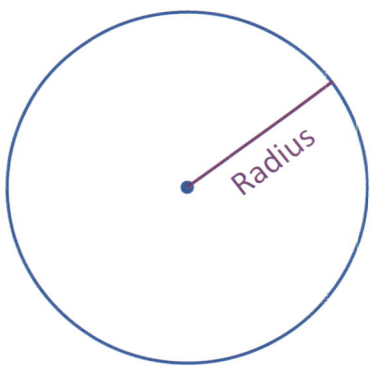

The area of a circle is proportional to the square of its radius. Once again the constant of proportionality is pi (π). The radius of a circle is half its diameter.

$$A_{circle} = \pi \times r^2$$

A circle with a radius of 10 cm therefore has an area of $\pi \times 10^2 \approx 314$ *cm²*.

Exercise 14.2

Calculate the area of the following shapes. Where necessary, give your answer correct to two decimal places. (Use $\pi = 3.14$ where necessary.)

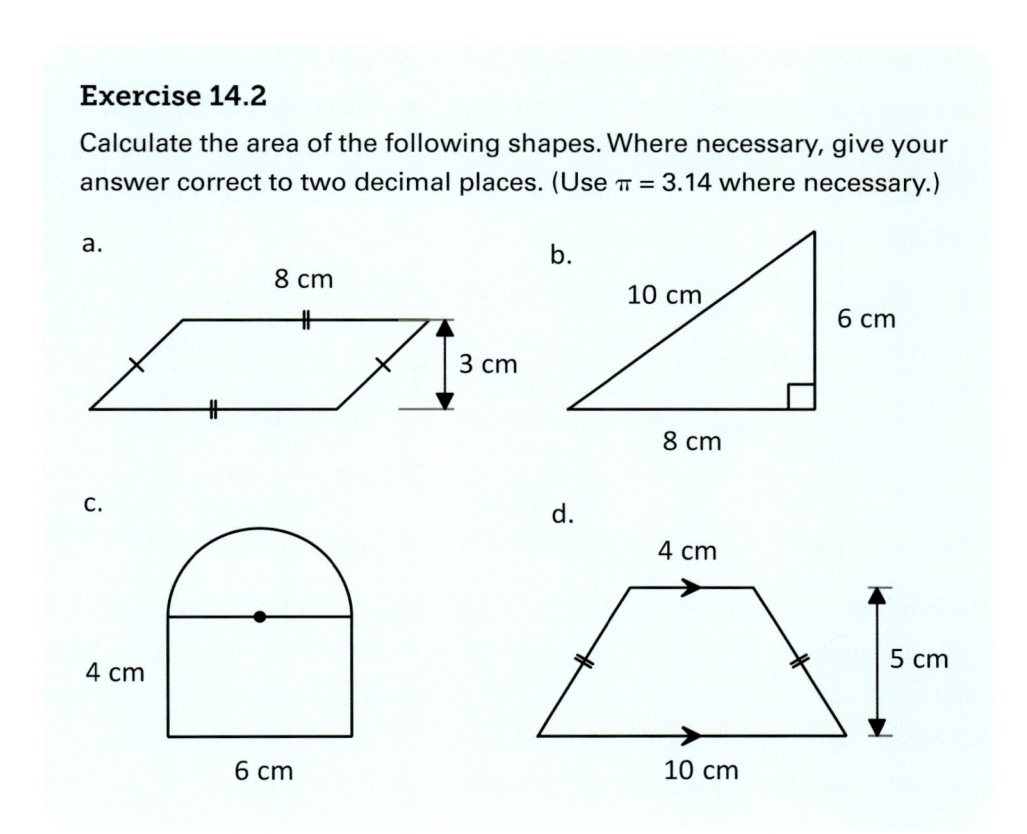

a.

8 cm

3 cm

b.

10 cm

6 cm

8 cm

c.

4 cm

6 cm

d.

4 cm

5 cm

10 cm

Polyhedra

Polyhedra are three-dimensional objects made by combining four or more polygons. The polygons are known as the **faces** of the object. A **vertex** (plural *vertices*) is a corner. **Edges** are line segments that join two vertices.

Both prisms and pyramids are types of polyhedra. Pictured opposite is a triangular pyramid. It has four triangular faces, four vertices and six edges.

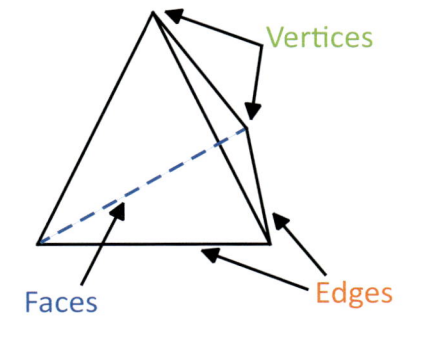

Vertices

Faces

Edges

Prisms are polyhedra consisting of two identical ends and flat sides. Prisms are named after the shape of the ends, such as triangular prisms, rectangular prisms and hexagonal prisms. All prisms have a constant cross-section, meaning that when sliced parallel to the ends, the cross-section remains the same size and shape (i.e., congruent).

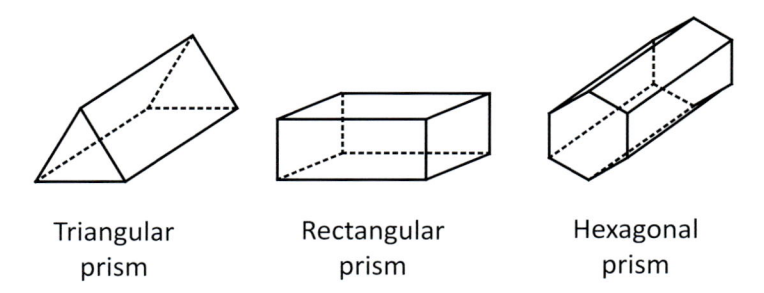

| Triangular prism | Rectangular prism | Hexagonal prism |

Pyramids are polyhedra with a polygonal base and triangular faces that taper to a point (known as the apex). Pyramids are named after the shape of the base, such as triangular pyramids, square-based pyramids, pentagonal pyramids and hexagonal pyramids. When sliced parallel to the base, the cross-section of a pyramid is similar to the shape of the base.

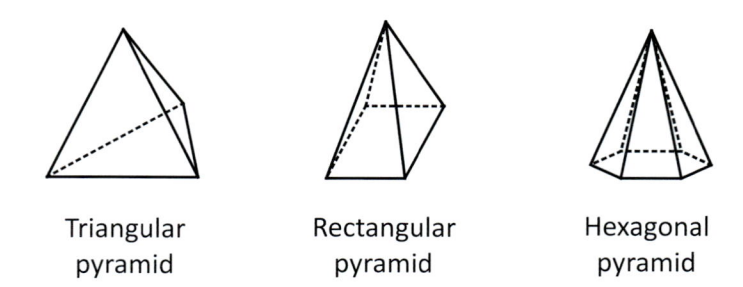

| Triangular pyramid | Rectangular pyramid | Hexagonal pyramid |

Volume of prisms and pyramids

The volume of any prism can be found by multiplying the area of its base by its height.

$$V_{prism} = A_{base} \times h$$

Example 1

A triangular prism with a base area of 20 cm^2 and height 10 cm has a volume of 200 cm^3.

The volume of any rectangular prism (sometimes called a cuboid) can be found by multiplying together the length, width and height.

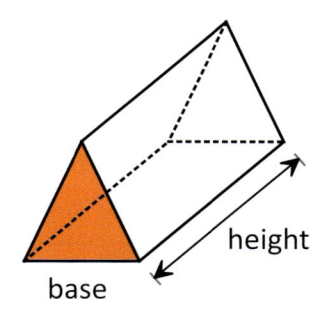

height

base

Example 2

A rectangular prism has dimensions 8 cm by 4 cm by 3 cm. Its volume is:

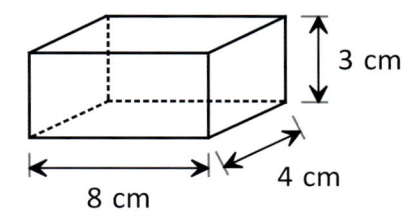

3 cm

4 cm

8 cm

$$V_{prism} = 8 \times 4 \times 3$$
$$= 96 \ cm^3$$

The volume of any pyramid is **one-third** the volume of the prism with the same size base and height.

Example 3

A rectangular-based pyramid is 8 cm long and 4 cm wide. It is 3 cm high. Its volume is found as follows:

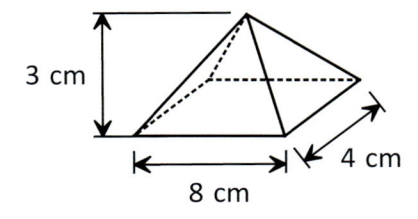

3 cm

4 cm

8 cm

$$V_{pyramid} = \frac{1}{3} \times (8 \times 4 \times 3)$$
$$= 32 \ cm^3$$

Australian Curriculum

The volume of a three-dimensional object is the amount of space that it occupies. Calculating the volume of rectangular prisms is introduced in Year 7 of the Australian Curriculum. Using formulas to calculate the volume of triangular prisms is introduced in Year 8, and cylinders in Year 9.

Exercise 14.3

Calculate the volume of the following prisms and pyramids. Where necessary, give your answer correct to two decimal places.

a.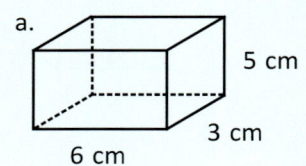

b.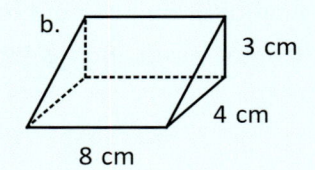

c.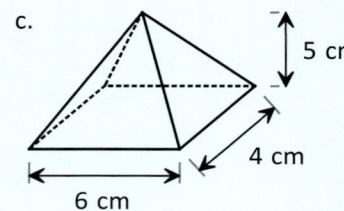

d.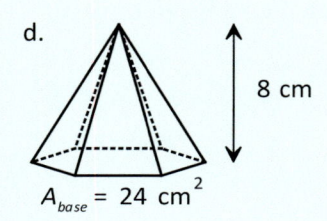

Volume of cylinders and cones

A cylinder is a solid with a flat base and a flat top joined by a single curved surface. The volume of a cylinder is found by multiplying the area of the circular base (i.e., πr^2) by its height:

$$V_{cylinder} = \pi r^2 \times h$$

Example 1

Calculate the volume of a cylinder with radius 10 cm and height 15 cm.

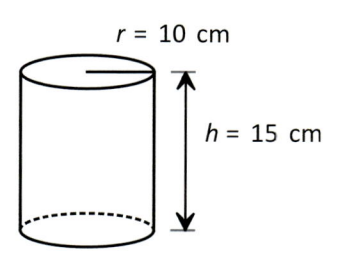

$$V_{cylinder} = \pi \times 10^2 \times 15$$
$$= 1\,500 \times \pi$$
$$\approx 4\,712.49 \; cm^3$$

A **cone** is a solid with a flat base and a single curved surface which tapers to a point. The volume of a cone is **one-third** the volume of the cylinder with the same base and height:

$$V_{cone} = \frac{1}{3} \times \pi r^2 h$$

Example 2

Calculate the volume of a cone with radius 5 cm and perpendicular height 8 cm.

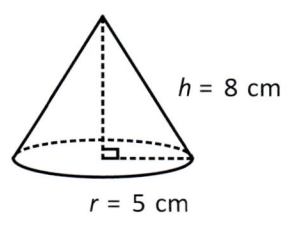

$h = 8$ cm

$r = 5$ cm

$$V_{cone} = \frac{1}{3} \times \pi \times 5^2 \times 8$$
$$= \frac{200\pi}{3}$$
$$\approx 209.44 \ cm^3$$

Exercise 14.4

Calculate the volume of the following cylinders and cones. Where necessary, give your answer correct to two decimal places. (Use $\pi = 3.14$.)

a.

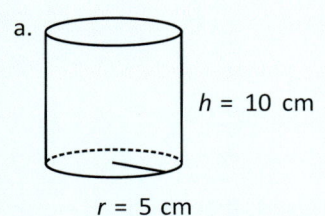

$h = 10$ cm

$r = 5$ cm

b.

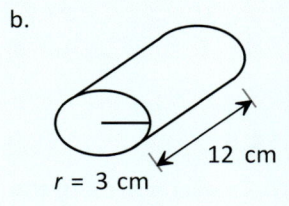

12 cm

$r = 3$ cm

c.

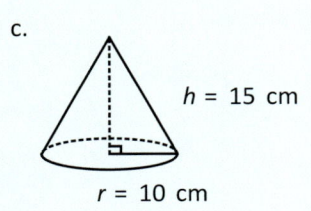

$h = 15$ cm

$r = 10$ cm

d.

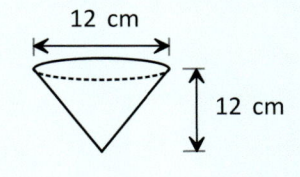

12 cm

12 cm

Nets and surface area

A net is a two-dimensional pattern that can be cut out and folded to make a three-dimensional object. For example:

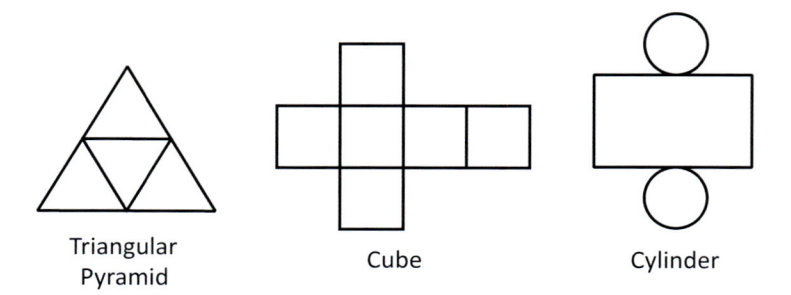

Triangular Pyramid   Cube   Cylinder

The **surface area** of a three-dimensional object is the total area of all of the surfaces (or faces) of the object. It can be found by adding together the areas of all of the shapes that make up the net of the solid.

Australian Curriculum

Students examine the nets of three-dimensional objects in the primary years, however, calculating the surface area of prisms and cylinders is not introduced until Year 9 of the Australian Curriculum.

Exercise 14.5

Calculate the surface area of the following objects. Where necessary, give your answer correct to two decimal places.

a.
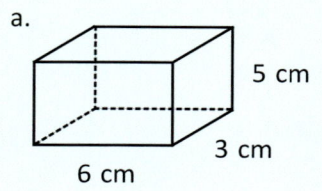
5 cm
3 cm
6 cm

b.
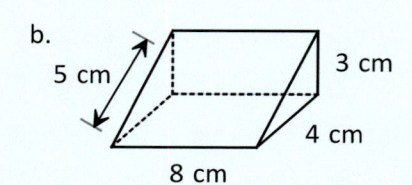
5 cm
3 cm
4 cm
8 cm

Transformations (2D)

A transformation involves changing either the size of a shape (i.e., dilation) or its orientation using flips (i.e., reflections), slides (i.e., translations) or turns (i.e., rotations). The original shape is referred to as the object and the transformed shape is referred to as the image. For example, when you look in a mirror the reflection that you see is your image.

Reflections

The reflection of a shape is its mirror image. The image appears to be the same size and distance from the mirror as the original object and is reversed. The mirror line is also referred to as a **line of reflective symmetry** (or simply line of symmetry). The mirror line can be horizontal, vertical or at any angle.

Some shapes (such as regular polygons) have many lines of symmetry. The examples below show that a parallelogram has no lines of reflective symmetry, while an isosceles triangle has one and a square has four. Lines of reflective symmetry can be found by folding a shape. A circle has an infinite degree of reflective symmetry.

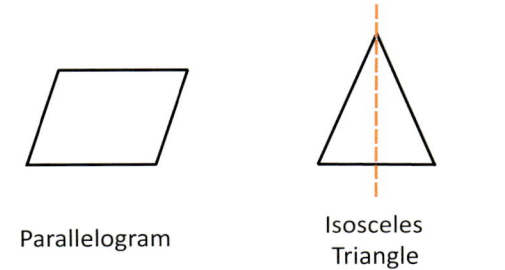

Parallelogram   Isosceles Triangle   Square

Activity

Draw the image when each object is reflected in the mirror line.

Activity

Draw in the lines of symmetry for each of the following shapes.

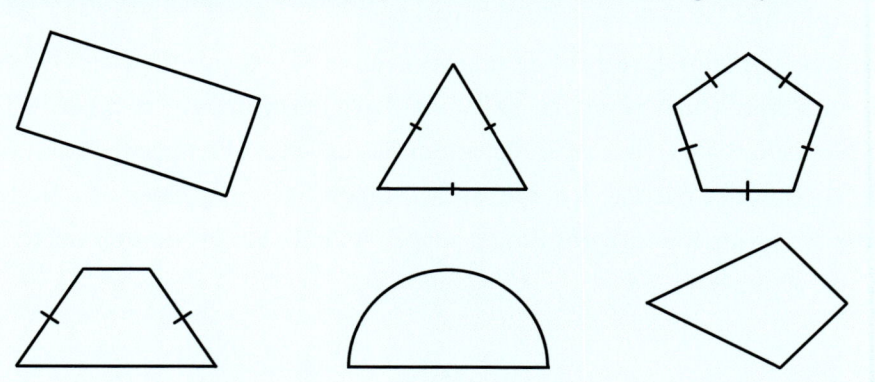

Rotations

When an object is rotated, the shape is turned through an angle either clockwise or anticlockwise to produce the image. Some shapes look the same after they have been

The fidget spinner has rotational symmetry order 3

rotated. The number of times the shape matches the object in a full turn is called the **order of rotational symmetry**.

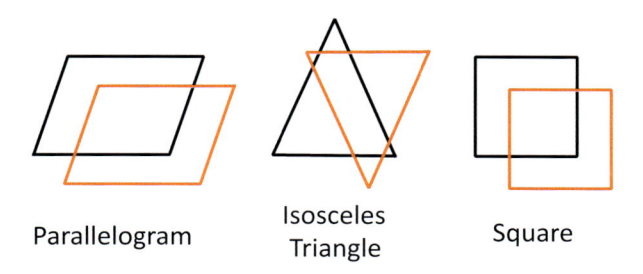

Parallelogram

Isosceles Triangle

Square

The images shown in red are the result of rotating each of the objects by 180°. The parallelogram has rotational symmetry order 2, since it looks the same when it is rotated by 180°. The isosceles triangle has rotational symmetry order 1, since it must be rotated a full 360° before it appears the same (in other words, an isosceles triangle does not have rotational symmetry). A square has rotational symmetry order 4, since it appears the same each time it is rotated by 90°.

Any regular polygon has an order of rotational symmetry equal to its number of sides. For example, a regular hexagon has an order of rotational symmetry of 6. A circle has an infinite degree of rotational symmetry.

Activity
Determine the order of rotational symmetry of the shapes.

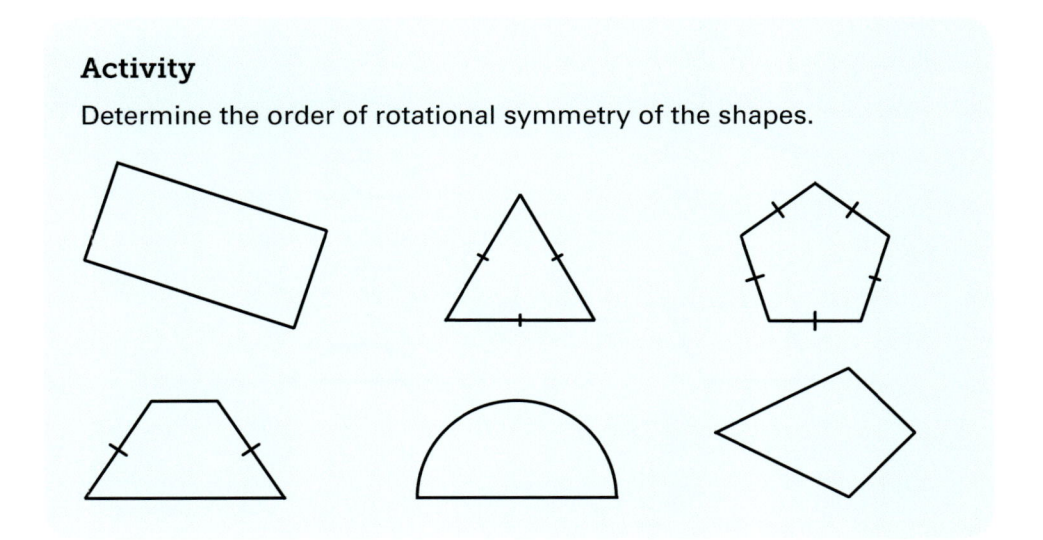

Translations

When a shape is moved in any direction, we say that it has been translated. The image is the same size and shape as the object. To translate a shape, every point in the shape must move the same distance in the same direction. Translations are sometimes described in terms of the horizontal and vertical distances moved. In the diagram below, the arrow has been moved 5 units to the right and 2 units upwards. This translation can be written as $\binom{5}{2}$. The oval has been moved 4 units to the right and 3 units down, which is written as $\binom{4}{-3}$.

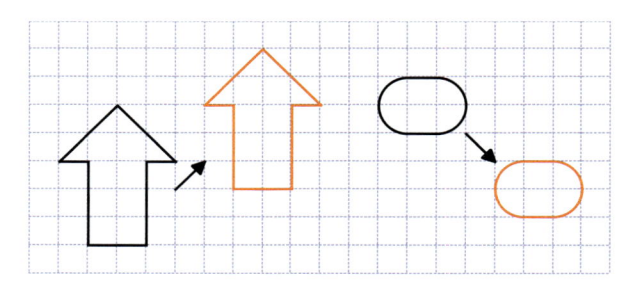

Activity

Translate the following shapes by the number of units indicated.

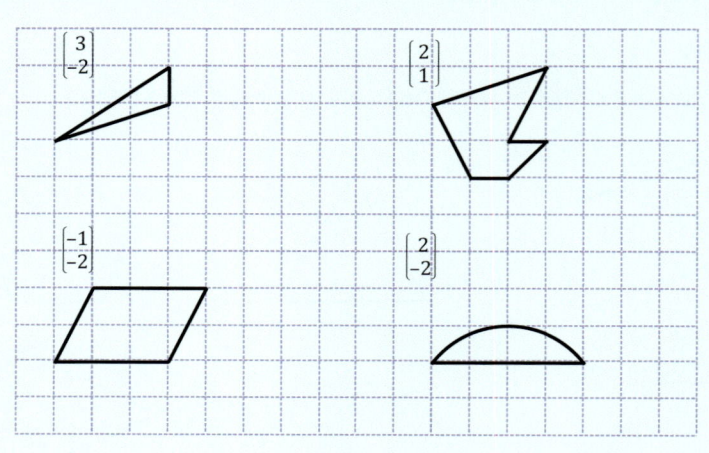

Australian Curriculum
In the later years of high school, students will be expected to write translations using vector notation i.e., $\begin{pmatrix} 5 \\ 2 \end{pmatrix}$.

Dilations

When an object is dilated, the image is either larger or smaller but remains the same shape. Mathematically speaking, we say that the object and the image are similar. When a shape is dilated, the length of each of its sides is multiplied by the scale factor. For example, using a scale factor of 2 means that each of the side lengths will double. Using a scale factor of $\frac{1}{2}$ means that each of the side lengths will halve.

To perform a dilation, we draw a line segment from the centre of dilation to each vertex. If performing an enlargement, we increase the length of the line segment by the scale factor. If performing a reduction, we decrease the length of the line segment by the scale factor. That is, for a scale factor of 2, each vertex is moved to twice the original distance from the centre of dilation. Regardless of the scale factor, the angles in the object and image remain the same size.

Example
Enlarge the green rectangle using the red centre of dilation. Use a scale factor of 3.

Solution

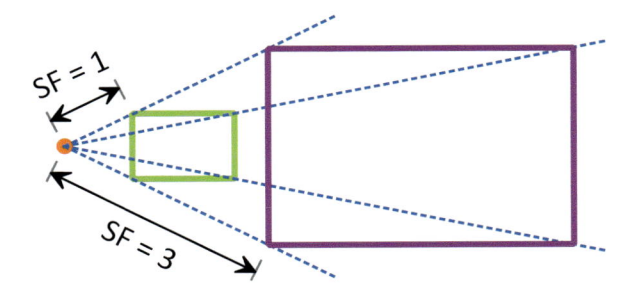

Each of the vertices in the image is three times as far from the centre of dilation as the corresponding vertex in the object.

Activity

Dilate the following shapes using the scale factor (SF) around the centre of dilation provided.

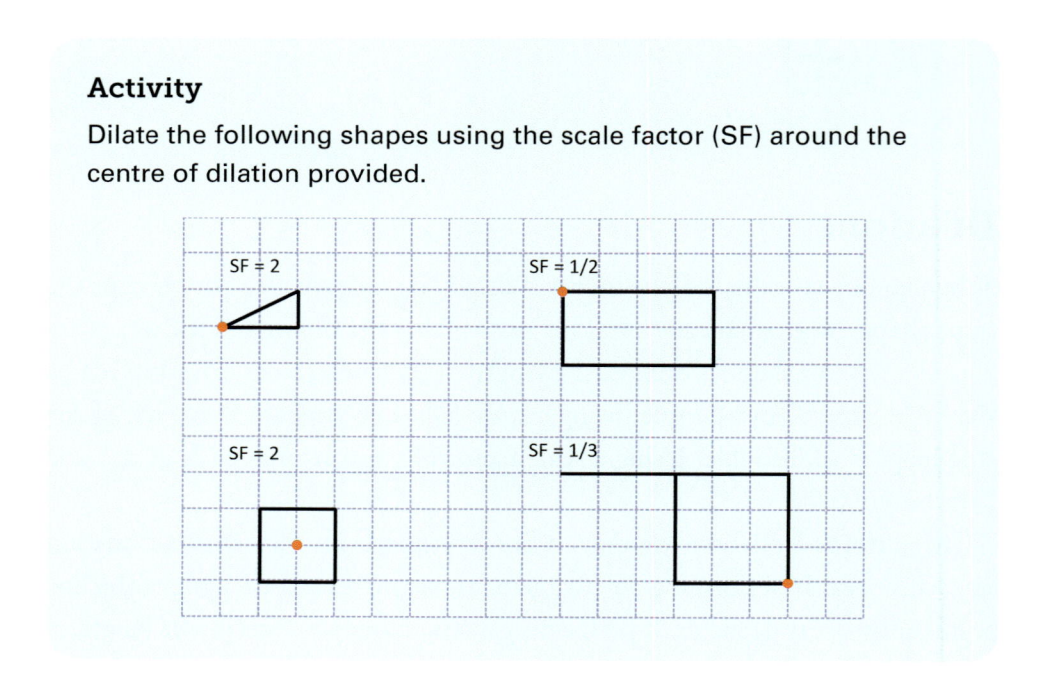

Location

Geometry is also concerned with locating points in two- and three-dimensional space. Students' early experiences with location are characterised by rather subjective concepts such as over, under, above, below, between, up, down, left, right, forwards and backwards. Over time students learn to give directions using the points of the compass (i.e., North, South, East and West), using grids and later coordinate systems.

Students first learn to specify locations using grid references, which describe the cells *between* the grid lines. In the example shown, we can see the yacht is in cell C4, while the

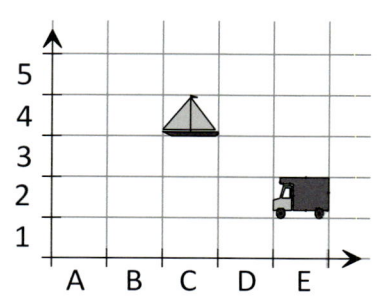

truck is in cell E2. Notice that here the numbers are being used as labels and not quantities.

Notice that a grid reference system cannot be used to show different positions within a cell. For this reason, specifying locations using a grid reference system is much less precise than using grid coordinates. A coordinate system is formed by placing two number lines at right angles to one another, as shown below.

A pair of numbers is used to indicate the exact position of a point. The first number is the horizontal coordinate and the second number is the vertical coordinate. In the example shown, the point (3, 2) is 3 units to the right and 2 units above the origin i.e., (0, 0).

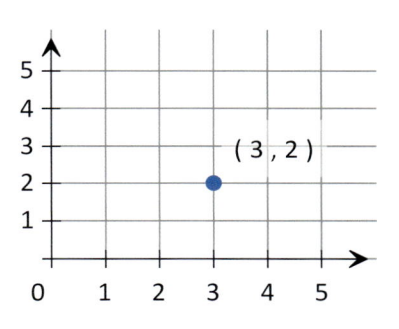

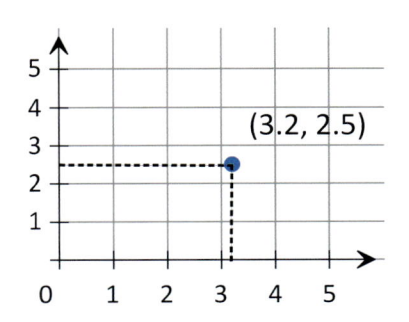

A coordinate system allows decimals (or fractions) to be used in order to specify more precise locations. For example, the point (3.2, 2.5) is slightly to the right and half a unit above the point (3, 2).

A set of coordinates is also known as an **ordered pair**. Changing the order of the coordinates changes the location of the point. The example shows the difference between the points (2, 5) and (5, 2).

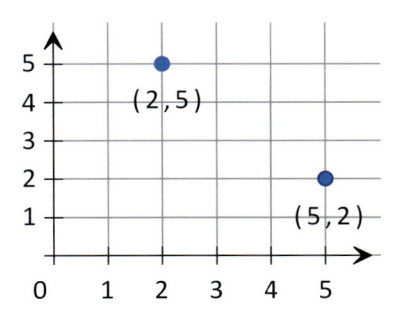

Chapter 15

Patterns and Algebra

A **pattern** is a repeated design or sequence that is arranged according to a rule. Almost every aspect of mathematics involves looking for, describing and using patterns in one way or another. For example, geometry involves looking for and describing patterns in shapes, while statistics involves looking for and describing patterns in data and using them to make predictions. In a similar manner, algebra is used to describe the general patterns that occur in arithmetic.

Repeating and growing patterns

Great emphasis is placed on the study of patterns during the primary years. Students' pre-algebra skills are developed by observing and recognising patterns of increasing complexity. Patterns can be divided into two broad categories: repeating patterns and growing patterns. In repeating patterns, the sequence remains the same each time the pattern repeats. For example:

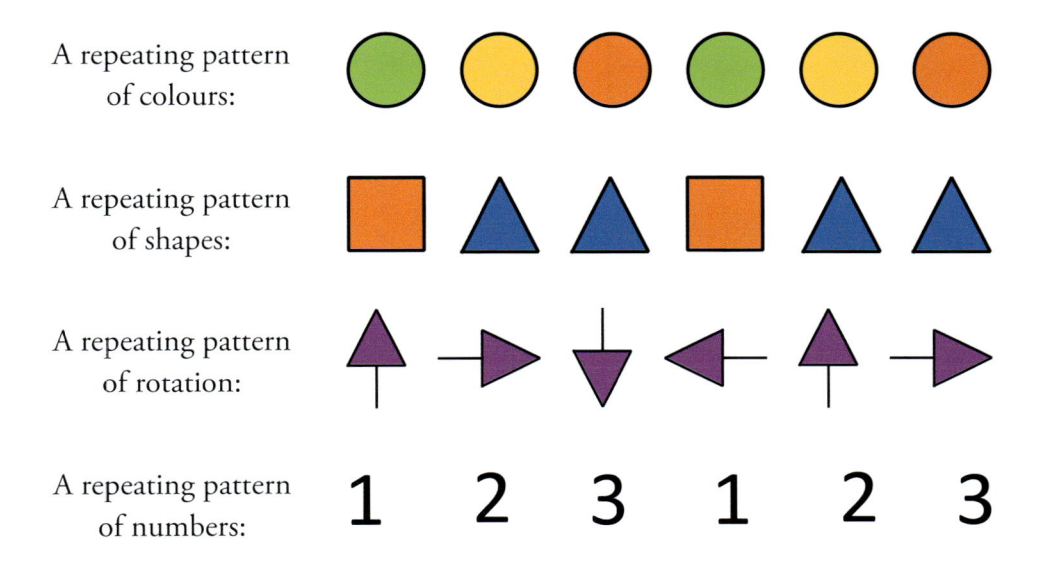

A repeating pattern of colours:

A repeating pattern of shapes:

A repeating pattern of rotation:

A repeating pattern of numbers:

In **growing** patterns, new elements are added each time the sequence repeats. For example:

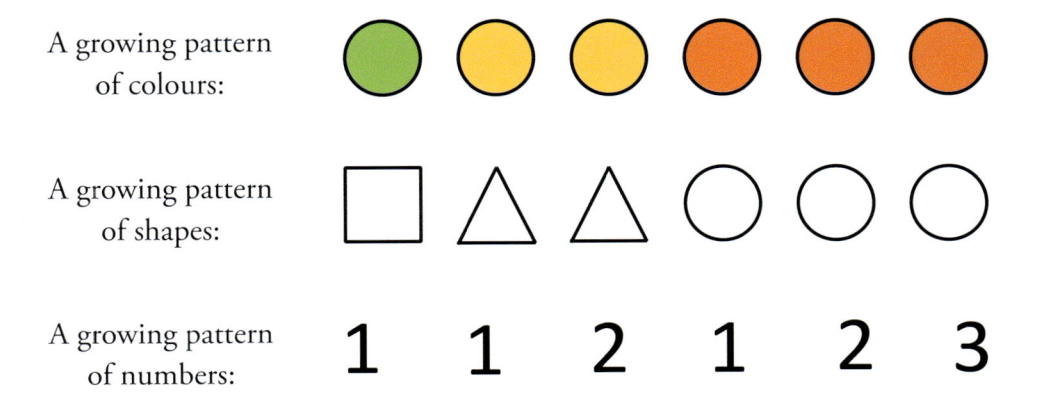

A growing pattern of colours:

A growing pattern of shapes:

A growing pattern of numbers:

Activity

Predict the next three terms in each of the following patterns.
Note: In some cases, there is more than one possible answer.

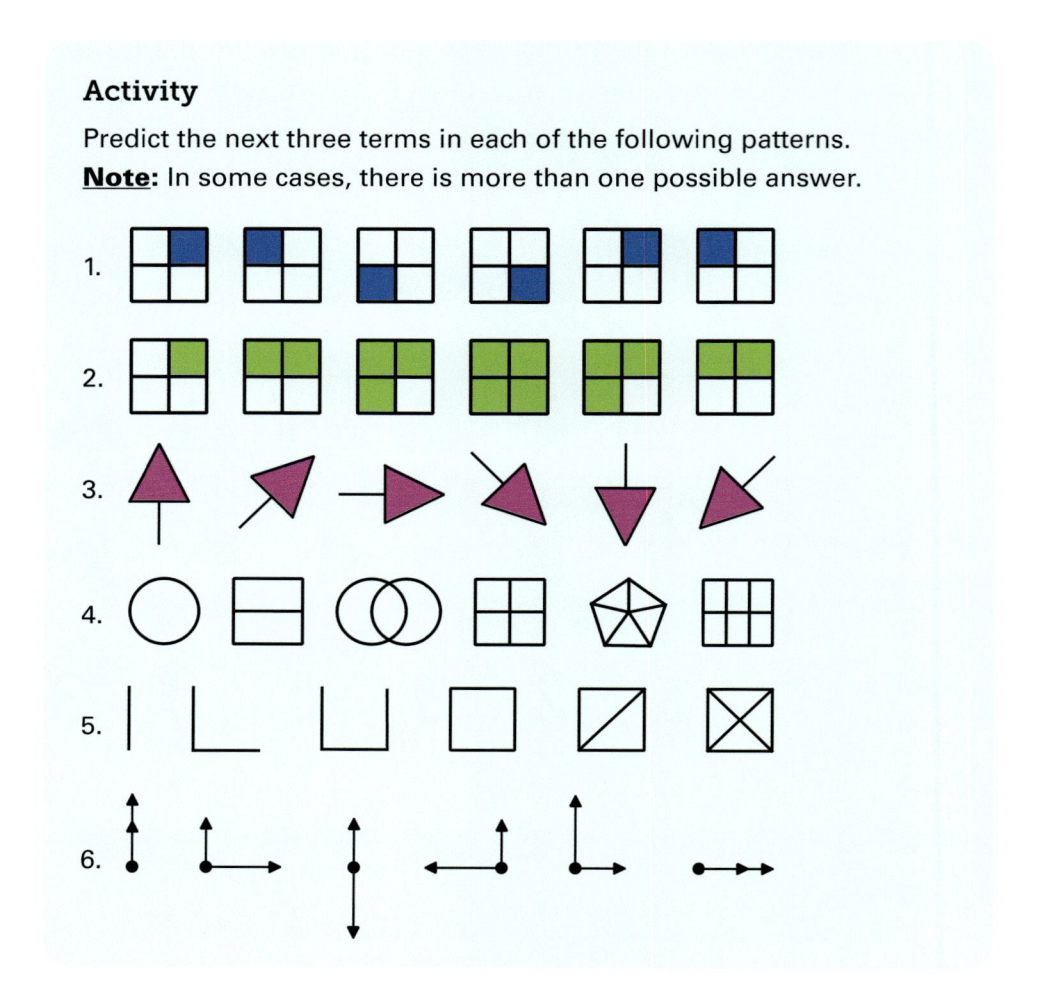

Number patterns

From an early age, students encounter number patterns in a range of contexts. The ability to recognise patterns is important in many aspects of everyday life. These include:

Number Pattern	Possible Contexts
Counting numbers	Birthdays
Odd and even numbers	House numbers in a street
Counting in sixes	Australian Rules Football, Cricket
Triangular numbers	Stacking cans, Ten-pin bowling
Square numbers	Board games e.g., chess, Tic-Tac-Toe

Odd and even numbers

A whole number is even if it is a multiple of 2 and odd if it is not. For example, 6 is even since $3 \times 2 = 6$, while 7 is not ($7 \div 2 = 3.5$). Even numbers can also be defined as numbers which are divisible by 2.

Note: While any number can be divided by 2, not all numbers are divisible by 2.

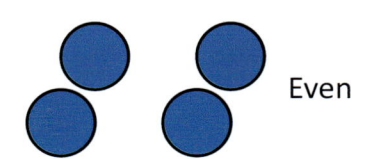

It is easy to distinguish odd and even numbers visually, since even quantities can be arranged into two equal piles. Odd and even numbers alternate on the number line. One more (or one less) than an odd number is an even number and vice versa.

Divisibility rules

Divisibility rules are patterns that can be used to work out whether a number is evenly divisible by another without the use of a calculator. Knowledge of the divisibility rules allows students to check the accuracy of calculations and increase the speed of mental arithmetic.

Number	Divisibility Rule	Examples
2	The number ends in 2, 4, 6, 8, or 0.	24, 36, 108
3	The digit sum of the number is divisible by 3.	1 + 2 + 3 = 6 so 123 is divisible by 3
4	The number formed by the last 2 digits is divisible by 4.	124, 1 288, 2 048
5	The number ends in 5 or 0.	95, 280, 3 125
6	The number is divisible by 2 and by 3.	216 is divisible by 6 since it is even and divisible by 3 (since 2 + 1 + 6 = 9)
8	The number formed by the last 3 digits is divisible by 8.	2 120 is divisible by 8 since 120 = 15 × 8
9	The digit sum of the number is divisible by 9.	5 + 6 + 7 = 18 so 567 is divisible by 9
10	The number ends in 0.	20, 50, 100, 12 500

Prime and composite numbers

The divisibility rules help us to more quickly determine whether numbers are prime or composite. A prime number is a number which is only divisible by 1 and itself. In other words, prime numbers have exactly two factors. On the other hand, a number is composite if it has more than two factors.

Number	Factors	Prime or Composite
1	1	1 is a special number. It is **neither** prime nor composite.
2	1, 2	2 is the only even prime since only $1 \times 2 = 2$
3	1, 3	3 is prime since only $1 \times 3 = 3$
4	1, 2, 4	4 is composite since $1 \times 4 = 4$ and $2 \times 2 = 4$
5	1, 5	5 is prime since only $1 \times 5 = 5$
6	1, 2, 3, 6	6 is composite since $1 \times 6 = 6$ and $2 \times 3 = 6$
7	1, 7	7 is prime since only $1 \times 7 = 7$
8	1, 2, 4, 8	8 is composite since $1 \times 8 = 8$ and $2 \times 4 = 8$
9	1, 3, 9	9 is composite since $1 \times 9 = 9$ and $3 \times 3 = 9$
10	1, 2, 5, 10	10 is composite since $1 \times 10 = 10$ and $2 \times 5 = 10$

One of the ways to determine a number is not prime is to show that it is divisible by a number other than 1 and itself. For example, any number that is divisible by 2 (excluding 2 itself) cannot be prime. Similarly, any number that is divisible by 3 (excluding 3 itself) cannot be prime. Notice that we do not need to test whether a number is divisible by 4, since any number that divides by 4 also divides by 2.

We can be sure that a number is prime if it is not divisible by any of the prime numbers up to the square root of the number.

Example

Determine whether 143 is prime or composite.

Solution

For 143 to be prime, it cannot be divisible by any of the prime numbers up to the square root of 143, which is approximately 11.96. Therefore we must test whether 143 is divisible by 2, 3, 5, 7 and 11. Notice that since 143 is not divisible by 2, there is no need to test if it is divisible by 6, 8, or 10 (which are multiples of 2). Since 143 is not divisible by 3 (since 1 + 4 + 3 = 8), there is also no need to test whether it is divisible by 9 (which is a multiple of 3).

Since 143 does not end in 5 or 0, it is not divisible by 5. Although there is no simple divisibility test for 7, we can use the fact that $20 \times 7 = 140$ and $21 \times 7 = 147$ to show that 143 is not a multiple of 7. Since $11 \times 11 = 121$ and $12 \times 11 = 132$, then $13 \times 11 = 143$. Therefore 11 is a factor of 143 and so 143 is composite.

Prime factoring

Any composite number can be decomposed into a unique set of prime factors. For example, we can write 30 as $2 \times 3 \times 5$ and 60 as $2 \times 2 \times 3 \times 5$ (which is also written as $2^2 \times 3 \times 5$). This approach can be illustrated using a factor tree.

A factor tree shows how a number can be written as the product of its prime factors. One possible factor tree for 72 is shown on the right. While other factor trees could be drawn, all produce the same set of prime factors. That is:

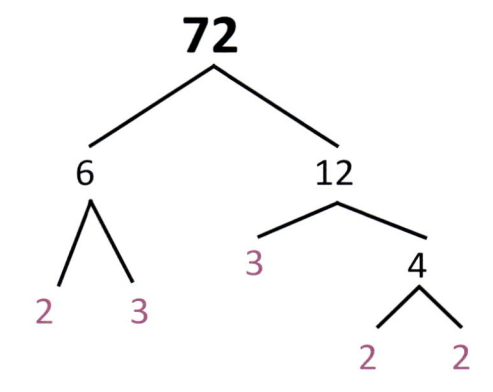

$$72 = 2 \times 2 \times 2 \times 3 \times 3$$
$$= 2^3 \times 3^2$$

Exercise 15.1

Write each of the following composite numbers as the product of their prime factors:

a. 88

b. 120

c. 400

d. 960

e. 1 080

f. 1 296

Highest common factor (HCF)

The highest common factor of two (or more) numbers can be found by first writing each number as the product of its prime factors. The highest common factor is found by multiplying together the prime factors that are common to all of the numbers.

Example

Find the highest common factor (HCF) of 90 and 220.

Solution

Create a factor tree for each of the numbers as shown below.

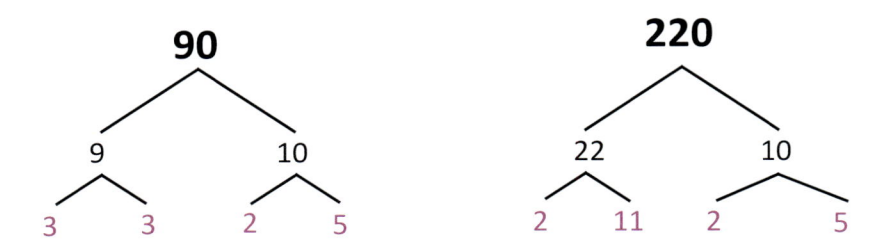

Since $90 = 2 \times 3^2 \times 5$ and $220 = 2^2 \times 5 \times 11$, the numbers share 2 and 5 as prime factors. Thus, the highest common factor (HCF) of 90 and 220 is $2 \times 5 = 10$.

Lowest common multiple (LCM)

The lowest common multiple of two (or more) numbers can be found by multiplying the numbers together and then dividing by the highest common factor.

Example

Find the lowest common multiple (LCM) of 90 and 220.

Solution

To find the LCM of 90 and 220, we calculate 90×220 and then divide by their HCF (i.e., 10). Therefore the *LCM* (90, 220) = 1 980.

Note: The LCM of a set of numbers **cannot** be smaller than the largest number in the set. The HCF of a set of numbers **cannot** be larger than the smallest number in the set.

Triangular numbers

The triangular numbers appear when objects are arranged in triangular shapes, such as the pins in a ten-pin bowling alley.

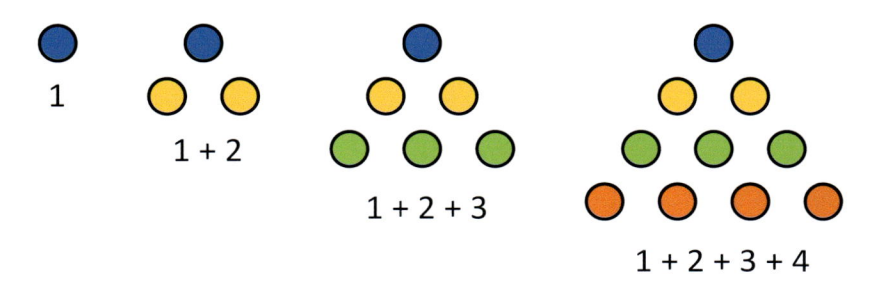

As can be seen in the diagram above, the first four triangular numbers are 1, 3, 6 and 10. Each new triangle is made by adding another larger row at the bottom and so the difference between each pair of triangular numbers increases by 1 each time. That is:

1 = 1

1 + 2 = 3

1 + 2 + 3 = 6

1 + 2 + 3 + 4 = 10 and so on.

Activity

What are the first ten triangular numbers?

Square numbers

The square numbers appear when objects are arranged in square arrays, such as the squares on a chessboard.

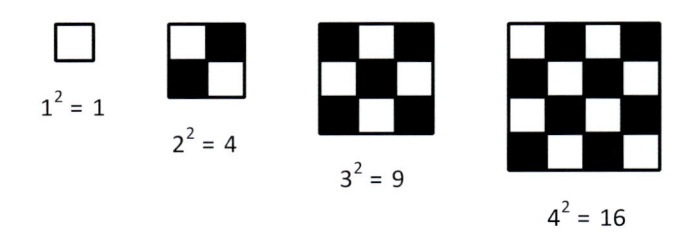

$1^2 = 1$

$2^2 = 4$

$3^2 = 9$

$4^2 = 16$

The first four positive square numbers are 1, 4, 9 and 16. Each square number is made by multiplying the counting numbers by themselves:

1 × 1 = 1
2 × 2 = 4
3 × 3 = 9
4 × 4 = 16 and so on.

Sequences

A sequence is a list of things (usually numbers) that occur in a particular order. The individual elements of a sequence are called terms. We use numbers (and later algebra) to label the terms of a sequence: $T_1, T_2, T_3, \ldots, T_n, T_{(n+1)}$.

Number sequences can be formed in many ways. These include:
- Adding or subtracting a constant number
  e.g., adding 2 to the previous term: 1, 3, 5, 7, 9, . . .
- Multiplying or dividing by a constant number
  e.g., doubling the previous term: 1, 2, 4, 8, 16, . . .
- A combination of the above
  e.g., double and add 1:   1, 3, 7, 15, 31, . . .
- Combining previous terms
  e.g., the Fibonacci sequence: 1, 1, 2, 3, 5, 8, . . .

- Adding an increasing amount
  e.g., the Triangular numbers: 1, 3, 6, 10, 15, . . .
- Multiplying numbers by themselves
  e.g., the Square numbers: 1, 4, 9, 16, 25, . . .

Number sequences can be formed in many ways. Understanding the different ways in which sequences can be formed may help us to work out the rule that describes a sequence. When looking at a sequence, we can first look to see if the rule involves adding or subtracting a constant number. We can do this by finding the differences between pairs of adjacent terms. If this does not appear to be the case, we might then look to see if the rule involves multiplying or dividing by a constant number. We do this by finding the ratio between adjacent terms (i.e., dividing one by the other). Although trial and error is an acceptable method, with practice, determining the rule for a sequence becomes much easier.

Type of Sequence	Examples
Adding a constant number	1, 4, 7, 10, 13, … (add 3) 7, 12, 17, 22, 27, … (add 5) 101, 201, 301, 401, 501, … (add 100) 2.6, 3.1, 3.6, 4.1, 4.6, … (add 0.5) $\frac{1}{2}, \frac{5}{8}, \frac{3}{4}, \frac{7}{8}, 1, \ldots$ (add $\frac{1}{8}$)
Subtracting a constant number	25.1, 24.8, 24.5, 24.2, 23.9, … (subtract 0.3) 72, 65, 58, 51, 44, … (subtract 7) $10, 9\frac{1}{2}, 9, 8\frac{1}{2}, 8, \ldots$ (subtract $\frac{1}{2}$)
Multiplying by a constant number	1, 2, 4, 8, 16, … (doubling) 3, 9, 27, 81, 243, … ($\times$ 3) 1 000, 100, 10, 1, 0.1, … ($\times$ 0.1) $1\frac{1}{2}, 2\frac{1}{4}, 3\frac{3}{8}, 5\frac{1}{16}, \ldots$ ($\times$ 1.5)

Type of Sequence	Examples
Dividing by a constant number	800, 400, 200, 100, 50, … (÷ 2) 0.125, 1.25, 12.5, 125, … (÷ 0.1) 10, 20, 40, 80, 160, … (÷ $\frac{1}{2}$)
A combination of the above	1, 3, 7, 15, 31, … (double and add 1) 7, 20, 59, 176, … (triple and take 1) 4, 2, −2, −10, −26, … (take 3 and double)
Combining terms	1, 1, 2, 3, 5, … (Fibonacci sequence) 1, 3, 4, 7, 11, … (Lucas sequence) 1, 2, 2, 4, 8, … (multiply previous 2 terms)
Adding an increasing amount	1, 3, 6, 10, 15, … (Triangular numbers) 1, 3, 7, 15, 31,… (add 2, 4, 8, etc.) 5, 10, 20, 35, 55, … (add 5, 10, 15, etc.)
Multiplying numbers by themselves	1, 4, 9, 16, 25, … (Square numbers) 1, 8, 27, 64, 125, … (Cubic numbers)

Exercise 15.2

Find the next three terms in each of the following sequences and explain the rule in each case.

1. 1, 2, 4, 7, 11, …

2. −16, −8, −4, −2, −1, …

3. $1\frac{1}{4}$, $2\frac{1}{2}$, $3\frac{3}{4}$, 5, $6\frac{1}{4}$, …

4. 1, 0.2, 0.03, 0.004, …

5. 100, 50, 25, 12.5, 6.25, …

6. 1, 4, 13, 40, 121, …

7. 0, 3, 8, 15, 24, …

8. 2, 2, 4, 6, 10, …

9. 1, 11, 121, 1331, 14 641, …

10. 1, 11, 111, 1 111, 11 111, …

11. 100, 99, 97, 94, 90, …

12. $\frac{1}{2}$, $\frac{5}{8}$, $\frac{3}{4}$, $\frac{7}{8}$, 1, …

13. 1, $1\frac{1}{2}$, $1\frac{3}{4}$, $1\frac{7}{8}$, $1\frac{15}{16}$, …

14. $\frac{1}{2}$, $\frac{2}{3}$, $\frac{3}{4}$, $\frac{4}{5}$, $\frac{5}{6}$, …

15. 5.5, 6.6, 7.7, 8.8, 9.9, …

Chapter 16

Statistics

Statistics is concerned with the collection, representation and interpretation of numerical data. We collect data for the purposes of identifying the patterns that exist within particular data sets (i.e., descriptive statistics) and to use these patterns to predict future trends (i.e., inferential statistics). This has important implications for business, education, agriculture, medicine and politics.

There are many ways in which data can be organised and presented, several of which are outlined in detail below. The use of statistical diagrams allows complex numerical information to be conveyed in a manner that is easy to read and interpret. The type of diagram used depends on whether the data being presented are numerical or categorical. Categorical data is classified or sorted into groups or categories. Numerical data can be further classified as discrete (in cases where only whole number values are possible, such as the number of children in a family) or continuous (in which case the data may take on any value within a given range, such as the length of a piece of string) (Brady & Winn, 2017).

Frequency tally

A frequency tally is often used prior to (or in conjunction with) the construction of a table or a graph. The standard method of tallying is grouping in fives, resembling four fingers and a thumb crossing the palm. Frequency tallies also allow categorical data to be collected as shown below.

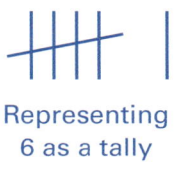

Representing 6 as a tally

Pets owned by class members

Pet Ownership	Tally	Frequency
Cat	ⅢⅡ	5
Dog	ⅢⅡ Ⅱ	7
Bird	Ⅲ	3
Fish	Ⅱ	2

Dot plots

A dot plot is a simple method of showing the frequency distribution of discrete numerical data. Here we see the frequency distribution of the scores obtained when a 6-sided die was rolled 12 times. From the dot plot, it is apparent that the most common score obtained was 6 (i.e., the mode) and the range of the scores was $6 - 1 = 5$.

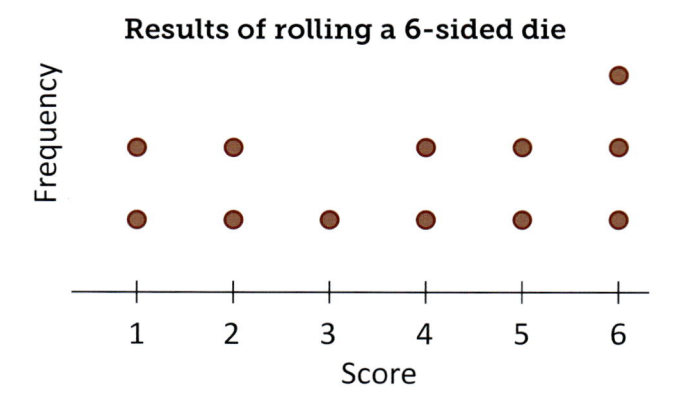

Picture graphs

A picture graph is used to represent the frequency distribution of categorical data. Here we see the frequency distribution of the types of pets owned by the students in a Year 5 class. From the picture graph, it is apparent that the most common type of pet was a dog (i.e., the mode).

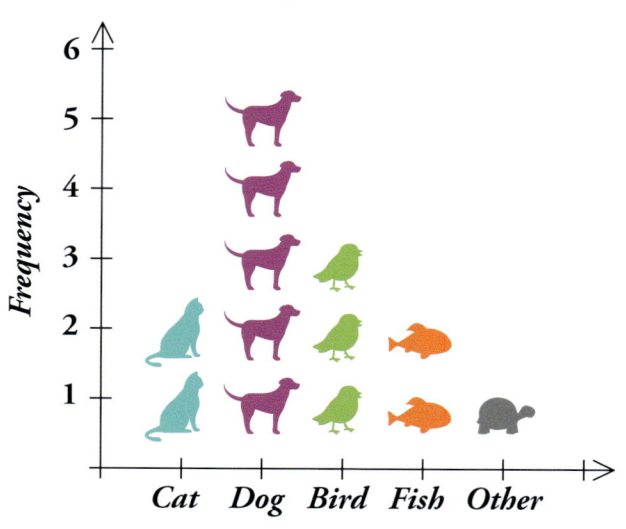

Picture graphs can be used to reinforce the idea of 1–1 correspondence with younger students (i.e., each picture represents one item), but can also be used to represent larger sets of data by incorporating a scale. For example, each transport symbol in the picture graph below may represent the journey of 10 students or 100 students, in which case a scale needs to be included.

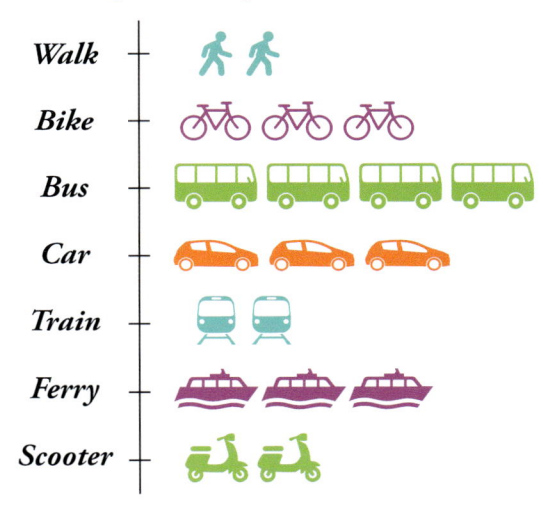

**Student modes of transport to school
(Each symbol represents 10 students)**

It is important that images in picture graphs are the same size in order to enable fair comparisons.

Bar and column graphs

Bar and column graphs are used to represent the frequency distributions of categorical data. For example, we could survey a class to find out their favourite type of fruit. For the set of data shown here, strawberries were the most popular type of fruit, with peaches being least popular. Notice that gaps have been left between the different types of fruit since they represent categorical data.

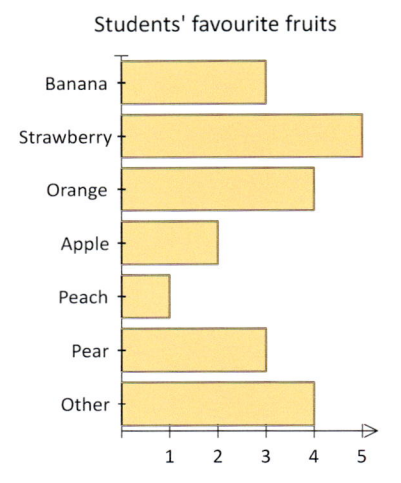

(This spacing distinguishes bar and column graphs from histograms which represent continuous data.)

Pie graphs

Pie graphs are a useful way of illustrating part–whole relationships, such as the individual components of a household budget. Pie graphs can be more difficult to construct since they require the measurement of angles. Teachers can assist by providing pre-prepared circles, allowing calculators to be used for calculations and/or the use of polar graph paper.

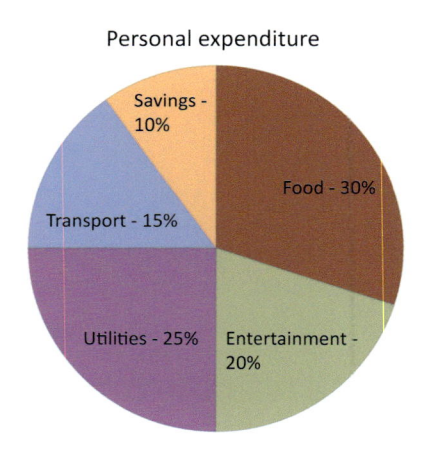

Histograms

Histograms are used to represent the frequency distribution of continuous numerical data. For example, the heights of the students in a class could be represented using a histogram. The number of students whose height falls within each class interval is shown on the vertical axis, while the class intervals are shown on the horizontal axis. Notice that unlike bar and column graphs, adjacent class intervals share a common boundary.

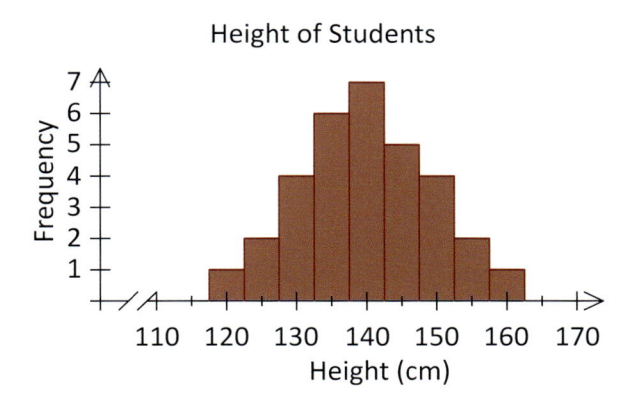

Line graphs

Line graphs are appropriate when the independent variable is continuous, such as measuring the height of a plant as it grows over time. In these instances, time is the independent variable and the other variable is the dependent variable. The example below shows how a student's bank balance changes over a period of three weeks (i.e., 21 days).

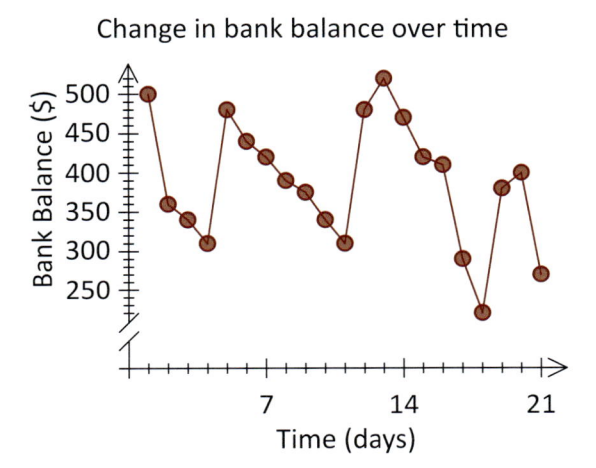

Stem and leaf plot

A stem and leaf plot provides a compact visual representation of a set of data. Each data value is broken into a stem (usually the first digit of the number) and a leaf (the last digit of the number). Data values are arranged in order, firstly by the value of the stem and then by the value of the leaf. Stem and leaf plots are useful for find the mode and identifying outliers.

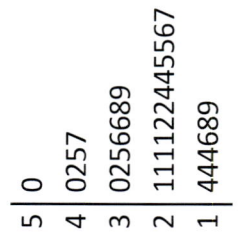

The stem and leaf plot shown here contains 30 data values, with the minimum score being 14 and the maximum score being 50. From the data shown, it is apparent that the majority of the scores were in the 20s and that the most frequently occurring score was 21. Notice that when the stem and leaf plot is turned sideways, it resembles a histogram.

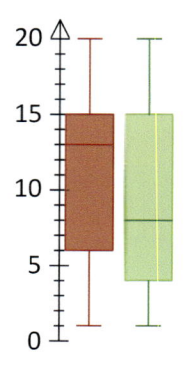

Box plot

Box plots (also known as box and whisker diagrams) provide a visual representation of a data set that shows:

- the maximum score
- the upper quartile
- the median (**not** the mean!)
- the lower quartile, and
- the minimum score.

The whiskers extend to the lowest and highest scores, while the box represents the middle 50% of scores for the group. The median is represented as a line across the box.

Side-by-side box plots allow two (or more) sets of data to be compared. The above diagram shows that both the red and green data sets had a range of 19.

The red data set had a much higher median than the green data set (13 vs 8) and the data values were more broadly spread in the green data set than in the red one.

Scatter graph

A scatter graph shows the relationship between two numerical variables. Scatter graphs can be used with discrete or continuous numerical variables. The scatter graph below shows the relationship between the height and shoe size of 30 male adults. Notice that it is possible to have people of different heights who wear the same sized shoes. The graph shows that, in general, taller males have larger feet.

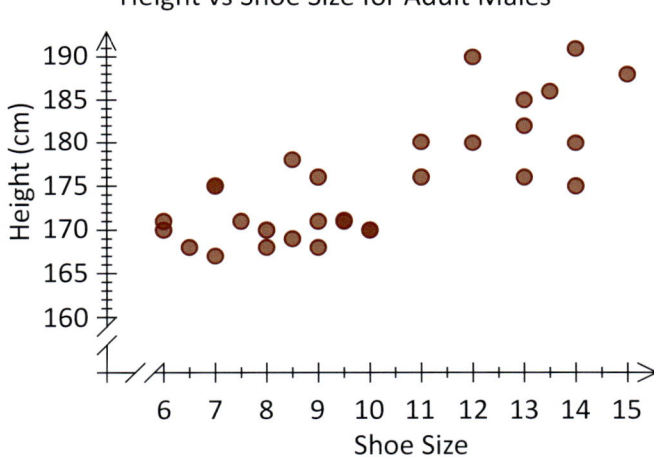

Descriptive statistics

In many instances it is useful to have a statistic that represents a typical or 'average' value in a data set. There are three common ways of determining an average value: the mean, the mode and the median. Collectively the mean, mode and median are known as measures of central tendency.

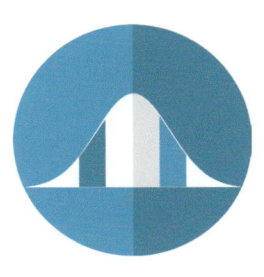

Of these, the mean is the most familiar type of average. It is calculated by adding all of the scores in a set of data and dividing by the number of scores. The mean is also known as the arithmetic average of a set of numbers.

$$Mean\ (\bar{x}) = \frac{Sum\ of\ Scores}{Number\ of\ Scores}$$

While the mean is a useful statistic, it is sensitive to the effect of outlying data values. The mean can also assume values that were not part of the original data set; for example, the average of 4, 5 and 9 is 6, which was not part of the original data set.

The mode is the score which occurs most frequently in the data set. For example, it would be of interest to shop owners to know the most common size of shoes sold when they are placing an order with their supplier. The mode is always one (or more) of the values in the original data set.

The median is the middle score when scores are arranged in order from lowest to highest. This can be illustrated by lining students up in order of height and finding the person who is in the middle (if there is one). This person is the median height. In situations where there is an even number of students, there will be no person exactly in the middle and so the median is found by taking the mean height of the two people closest to the middle position. Another way to find the median of a set of data is to systematically cross off the lowest and highest scores of a set of data until just one (or two) middle scores remain.

Example

The heights of four students are 155 cm, 168 cm, 149 cm and 145 cm. The median height is found by first arranging the scores in order (i.e., 145, 149, 155, 168) and then crossing off the highest and lowest scores until only the middle score(s) remain (i.e., 149 and 155). If there are two middle scores, the median is found by taking the mean of these scores i.e., 152 cm.

Australian Curriculum
Students learn to read, interpret and construct various graphs in the primary years, but they do not begin to construct their own stem and leaf plots until Year 7 and box plots in Year 10.

  A great deal of NAPLAN data is provided in the form of box plots so, as a teacher, you will be expected to know this content.

Exercise 16.1

1. Find the mean of the following sets of data:
    a. 168 cm, 170 cm, 171 cm, 165 cm, 183 cm
    b. 480 g, 360 g, 120 g, 250 g, 375 g
    c. 2.4, 3.8, 7.1, 11.3, 16.4, 12.9
    d. $150 000, $200 000, $250 000, $10 000 000
2. Find the mode of the following sets of data:
    a. 1, 3, 4, 6, 8, 0, 1, 7, 6, 5, 3, 2, 1, 1, 2, 2, 6, 4
    b. 0.1, 0.3, 0.7, 0.5, 0.4, 0.5, 0.6, 0.4, 0.2, 0.8, 0.4, 0.5
    c. 14, 20, 14, 16, 18, 12, 14, 17, 19, 13, 12
    d. 1, 2, 1, 1, 2, 1, 2, 2, 1, 2, 1, 2, 2
3. Find the median of the following sets of data:
    a. 7, 9, 11, 13, 13, 1,3, 5, 7, 9, 11, 10
    b. 2, 4, 6, 8, 6, 4, 6, 8, 6, 4, 2
    c. 1, 1, 2, 3, 4, 4, 5, 6, 7, 9, 9, 9
    d. 0.1, 0.4, 0.2, 0.2, 0.5, 0.1, 0.5, 0.3

Chapter 17

Probability

Events that cannot be predicted with certainty are known as **chance events**. The **probability** of an event is the likelihood that the event will occur. The likelihood of a chance event occurring can be assigned a value from 0 to 1, where 0 means that the event is impossible and 1 means that the event is certain to occur. The probability of an event occurring is normally written as either a fraction, decimal or percentage and can be located on a probability scale such as the one shown below.

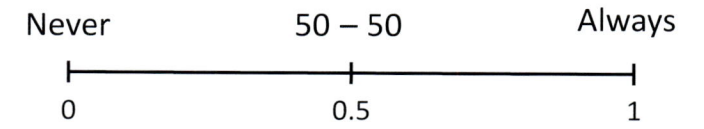

There are two ways in which the probability of an event can be calculated. One way is to use data obtained from a chance experiment (or simulation) to predict the probability of the event in the future. The other approach is to systematically list all of the possible outcomes of the chance experiment in order to work out the probability that the desired event will occur.

Experimental probability

When we toss a fair coin, we know that the probability of getting a 'head' is $\frac{1}{2}$ or 0.5. But we also know that when we toss the coin twice, there is no guarantee that we will obtain one 'head' and one 'tail', since the toss of a coin is a chance event. Similarly, if you were to toss a coin 100 times, although you would expect to get 'heads' approximately half of the time, you would not be surprised if you did not get exactly 50 'heads' and 50 'tails'.

Experimental probability describes what *actually* happens when a chance experiment is conducted (Brady & Winn, 2017). If we were to toss a coin 100 times and obtain 52 'heads' and 48 'tails', the experimental probability of 'heads' would be $\frac{52}{100}$ or 0.52. Since the experimental probability is based on empirical data, there is no guarantee that if we were to repeat the experiment we would obtain the same (or even similar) results. That is one reason why predictions such as economic forecasts and weather forecasts actually differ from reality.

Exercise 17.1

David recorded the colours of 20 cars in the school carpark as shown:

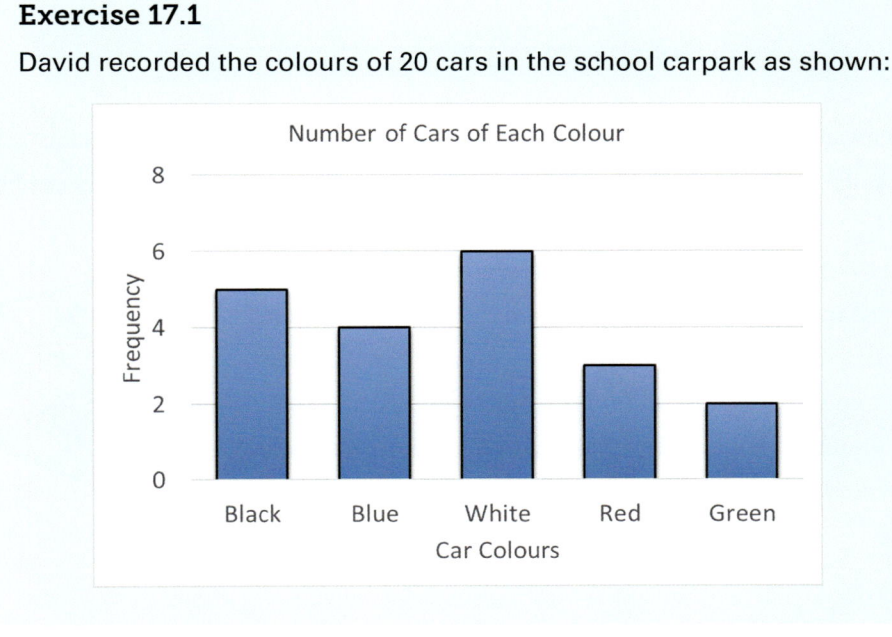

> **Exercise 17.1** cont
>
> Determine the experimental probability that a randomly chosen car is:
>
> a. black
> b. red or green
> c. yellow
> d. any colour except green

Theoretical probability

Theoretical probability is what you *expect* to happen when a chance experiment is conducted (Brady & Winn, 2017). In order to calculate theoretical probability, we must first determine all of the possible outcomes of the chance experiment. This is known as the sample space. The theoretical probability of an event, written as $P(E)$, is calculated as follows:

$$P(E) = \frac{Number\ of\ ways\ an\ event\ can\ occur}{Total\ number\ of\ possible\ outcomes}$$

If we were to toss a coin 100 times, we would expect to obtain 50 'heads' and 50 'tails' since there are only two possible outcomes, each of which is equally likely.

For the five-sided spinner shown, we would expect the chance of spinning a '2' to be 1 in 5 since there are five equally likely outcomes. If we were to repeat the experiment a total of 50 times, we would expect to obtain each number 10 times. This is known as the expected value.

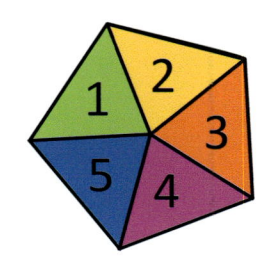

Exercise 17.2

This spinner consists of eight equal-sized regions. If the spinner is spun, what is the theoretical probability that a spin produces:

a. blue
b. red or green
c. yellow
d. any colour except green

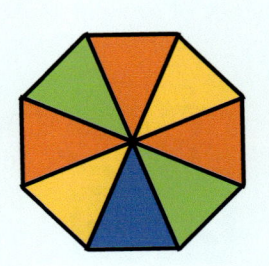

Sample spaces

A sample space is a representation of all of the possible outcomes of a chance experiment. Sample spaces can take the form of a list, table, grid or tree diagram. It is important to be systematic when preparing a sample space, since all of the possible outcomes must be identified if the theoretical probability is to be calculated accurately.

For example, consider the sample space for rolling one fair 6-sided die. There are six possible outcomes: {1, 2, 3, 4, 5, 6}. The chance of each outcome occurring is $\frac{1}{6}$ since all outcomes are equally likely. That is: $P(1) = P(2) = P(3) = P(4) = P(5) = P(6) = \frac{1}{6}$.

Consider the situation in which a die is rolled i.e., {1, 2, 3, 4, 5, 6} and a coin is tossed i.e., {H, T}. In this situation, the possible outcomes of this experiment can be represented using a table.

		Result of die roll				
	1	2	3	4	5	6
Coin toss *H*	{H, 1}	{H, 2}	{H, 3}	{H, 4}	{H, 5}	{H, 6}
T	{T, 1}	{T, 2}	{T, 3}	{T, 4}	{T, 5}	{T, 6}

The table contains a total of $6 \times 2 = 12$ possible outcomes. Therefore, the chance of any particular outcome is $\frac{1}{12}$.

Australian Curriculum

In the primary years students investigate chance outcomes and learn that probabilities range from 0 to 1. They compare frequencies across experiments with expected frequencies, however, terminology such as experimental probability and theoretical probability is not introduced until the high school years. Sample spaces are introduced in Year 7 of the Australian Curriculum.

Exercise 17.3

Use a list or table to represent the sample space for the following events.

a. A ten-sided die (which has numbers 0–9) is rolled.
b. A letter is chosen at random from the set of vowels.
c. A month of the year is chosen at random.
d. A day of the week is chosen at random.
e. A number is chosen at random from the factors of 12.
f. One of the Australian states is chosen at random.

Tree diagrams

Tree diagrams provide a useful way of representing multi-stage probability experiments. Imagine that two cards are selected from a standard deck of 52 playing cards containing 26 red cards (i.e., hearts and diamonds) and 26 black cards (i.e., clubs and spades). The first two branches of the tree (on the left) show the possible outcomes of the first stage of the experiment—the first card selected is either red or black. Regardless of whether the first card selected is red or black, the second card selected may also be red or black, giving four possible outcomes of the experiment: {*RR, RB, BR, BB*}.

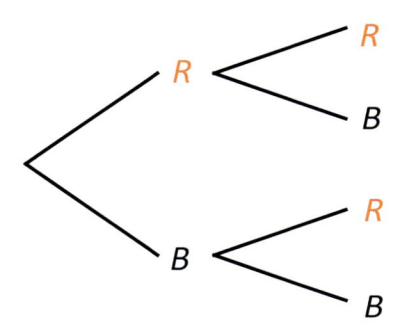

Independent events

Two events are independent when the occurrence (or non-occurrence) of the first event has no bearing on the chance of occurrence of the other. For example, imagine a coin is tossed three times in a row. Each time the coin is tossed, there are just two possible outcomes i.e., {*H, T*}. The sample space lists all of the possible outcomes for the experiment.

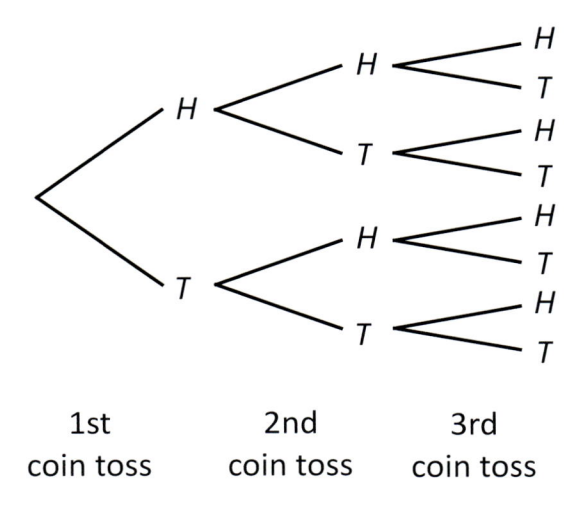

The first two branches of the tree diagram (on the left) show the possible outcomes of the first stage of the experiment i.e., 'heads' or 'tails'. Each of these outcomes is followed by two possible outcomes i.e., 'heads' or 'tails', accounting for the middle four branches. Each of these four outcomes (i.e., *HH*, *HT*, *TH*, or *TT*) is followed by two possible outcomes i.e., 'heads' or 'tails', giving a total of eight possible outcomes.

1st toss	2nd toss	3rd toss
H	H	H
H	H	T
H	T	H
H	T	T
T	H	H
T	H	T
T	T	H
T	T	T

Exercise 17.4

A family is dining in an Italian restaurant. They have chosen a three course set menu which allows them two choices of entrée (antipasto or bruschetta), three choices of main meal (fish, chicken or beef) and two choices of dessert (gelato or tiramisu). Construct a tree diagram showing all possible meal combinations.

Dependent events

In some cases, the events that occur in the first stage of a multi-stage probability experiment can affect the probability of the events that may occur in subsequent stages. For example, consider a tin can containing four marbles: one red, one black, one green and one white. If the first stage of the experiment involves drawing out one marble at random, the sample space is {*red, black, green, white*}. The chance of any individual marble being drawn is 1 in 4. If a second marble is then drawn *without* replacing the first, there are now only three marbles left to choose from, increasing the chance that each will be drawn to 1 in 3. This can be illustrated using a tree diagram:

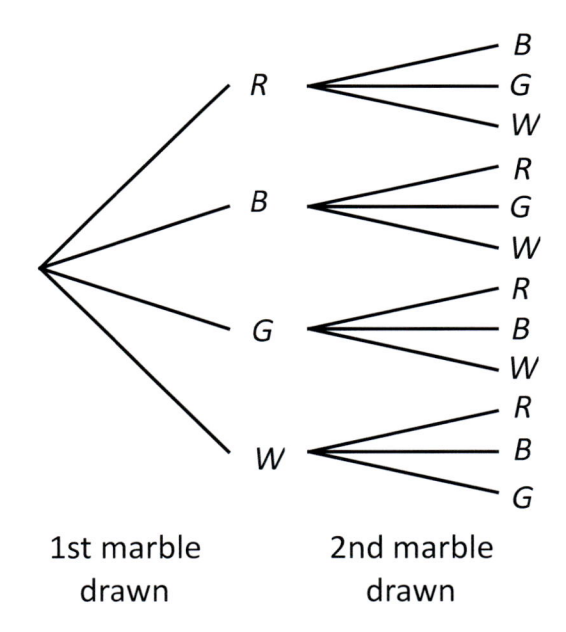

Notice that if the red marble is selected in the first stage of the experiment, it can no longer be selected in the second stage. Therefore, the probability that any marble is selected in the second stage is dependent on the probability it was selected in the first. The same is true regardless of the colour selected in the first stage.

Exercise 17.5

Determine whether the outcomes of the following chance experiments are dependent or independent.
a. Two coins are tossed at the same time.
b. Two coins are tossed, one after the other.
c. Two cards are selected from a deck without being replaced.
d. Winning both Monday Lotto and Saturday Lotto.
e. My first and second child will both be boys.

Mutually exclusive events

Two events are **mutually exclusive** if the occurrence of one event precludes the occurrence of the other. For example, there is a 1 in 365 chance that today is your birthday and a 364 in 365 chance that it is not. These events are mutually exclusive since it is either your birthday today or it is not.

Exercise 17.6

Determine which of the following pairs of events are mutually exclusive.
a. It rains on a windy day.
b. A letter is both a vowel and a consonant.
c. A number is prime and composite.
d. A red card is also a picture card (i.e., Jack, Queen or King).
e. A number card is also a picture card.
f. Rolling a die produces an even and odd number.

Australian Curriculum
The concept of mutually exclusive events is introduced in Year 8 of the Australian Curriculum.

Acknowledgements

This project could not have been completed without the constant support and encouragement of our families, friends and colleagues.

John West and Fiona Budgen
February 2019

Glossary

Language plays a crucial role in the teaching and learning of mathematical concepts. There is an extensive range of mathematical terminology and the way in which terms are used in mathematical contexts often differs from their conventional usage. Recognising this, the Australian Curriculum, Assessment and Reporting Authority (ACARA) has included a comprehensive glossary of mathematical terms in the Australian Curriculum: Mathematics, which is available from www.australiancurriculum.edu.au/f-10-curriculum/mathematics/glossary/.

The glossary presented here focuses on the mathematical terminology that is used in the primary years of schooling. Understanding the content that you teach (AITSL Standard 2) includes the ability to explain mathematical concepts and ideas using the appropriate mathematical terminology. Obviously, there is a great deal of mathematical vocabulary and it is important not to overwhelm students.

It is suggested that wherever possible teachers model the use of appropriate terminology to communicate mathematical ideas. For example, the teacher's lesson plan might identify several key terms to be explored in a lesson (or series of lessons) on fractions (such as numerator, denominator and equivalent). The precision with which terminology is used will also depend on the developmental level of the students.

Algebra

In the primary years, algebra focuses on the patterns and properties of numbers.

Algorithms

An algorithm is a step-by-step procedure for performing a calculation. Algorithms can be performed without any understanding of the numbers involved in a calculation.

Angles

Angles are measures of the amount of turn. An angle consists of two rays that meet at a vertex. Angles are classified into categories based on their size. The Greek letter theta (θ) is often used to denote an angle.

Acute	θ = between 0° and 90°
Right	θ = 90°
Obtuse	θ = between 90° and 180°
Straight	θ = 180°
Reflex	θ = between 180° and 360°
Revolution	θ = 360°

Area

Area measures the number of units required to cover a surface. Area is measured in square units, such as square metres or square centimetres.

Array

A rectangular arrangement of objects in equal rows or columns. Arrays are often used to look at multiplicative relationships. The example shows a 3×4 array.

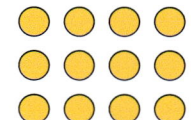

Associative property

The way in which numbers are associated (or grouped) in an addition or multiplication does not affect the result. For example, $(3 + 4) + 5 = 3 + (4 + 5)$ and $(3 \times 4) \times 5 = 3 \times (4 \times 5)$. Subtraction and division are not associative.

Attribute

Attributes are properties or characteristics of objects that allow them to be classified or measured. Attributes that we measure include length, area, volume, capacity, mass, time, temperature and angles.

Average

Three types of average are used at primary school level. These are the *mean*, the *mode* and the *median*.

Axes

Two number lines drawn at right angles to define a *coordinate system*. Typically, the horizontal axis is used for the independent variable (x) and the vertical axis is used for the dependent variable (y).

Base 10

The number system that uses the digits 1, 2, 3, 4, 5, 6, 7, 8, 9 and 0 to represent numbers.

Bivariate data

Bivariate data consists of two variables. For example, the height and arm spans of the students in your class.

Box and whisker plot/Box plot

A box-and-whisker plot is a graphical display that shows the minimum and maximum values in a set of data, the median and the upper and lower quartiles.

Capacity

The capacity of a container refers to the amount that the container can hold. Capacity is generally measured in litres.

Cardinal number

A *whole number* which tells us how many objects are in a set.

Cartesian plane

The Cartesian plane describes the location of any point in a plane using an ordered pair of numbers, called coordinates.

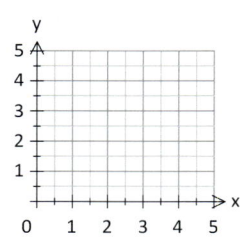

Categorical variable

A categorical variable is a variable whose values are categories. For example, the day of the week or the month in which you were born.

Census

A census is a survey of an entire population.

Chance event

A situation in which the outcome is not certain. For example, the toss of a coin or the roll of a die.

Circle

A circle is the set of all points that are equidistant from the centre.

Circumference

The circumference refers to the perimeter of a circle. Circumference is calculated using the formula $C = \pi \times d$.

Column graph

A column graph is used to display categorical data. The height of a column corresponds to the frequency of the category. A side-by-side column graph allows two sets of data to be compared.

Common factor

A common factor of a set of numbers or expressions is a factor of each element of that set. For example, 6 is a common factor of the set consisting of 24, 54 and 66.

Common multiple
A multiple that is shared by two or more numbers. For example, 12 is a common multiple of 2, 3 and 6.

Commutative property
The order in which two numbers are added or multiplied does not affect the result. For example, $4 + 5 = 5 + 4$ and $4 \times 3 = 3 \times 4$. Subtraction and division are not commutative.

Composite number
A composite number is a number that has more than two factors. Composite numbers are not prime. For example, 12 is composite since it has 1, 2, 3, 4, 6 and 12 as factors.

Computation
Operations on numbers that are used to calculate a result.

Concave
A shape that has at least one interior angle that is greater than 180°. A concave shape has an indentation.

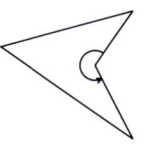

Cone
A cone is a solid with a circular base rising to a point (called the *vertex*) joined by an oblique surface.

Congruent
Two figures are congruent if they are the same shape and the same size.

Conjecture
A mathematical statement whose truth has yet to be established.

Continuous numerical data
Continuous numerical data may take on any value within a particular interval. For example, the length of a piece of string.

Convex
A shape whose interior angles are all less than 180°. All regular polygons are convex.

Coordinate

A coordinate is one of two values in an ordered pair that describe a location in the Cartesian plane.

Coordinate system

A system used to assign numbers to locate points in space.

Counting numbers

Counting numbers are the positive *integers* (i.e., 1, 2, 3, …).

Counting on

Counting on is a technique for solving simple addition and subtraction problems using a number line.

Cylinder

A cylinder is a solid with circular ends and a constant cross-section. Each horizontal cross-section is a circle with the same radius.

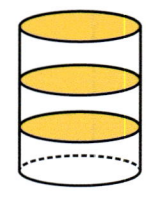

Data

Information collected in a systematic manner.

Data display

There are various methods in which data may be displayed. The method used will depend on the type of data being presented.

Decimal

A decimal is a number expressed in the Base 10 system such that the value of each *digit* is 10 times the value of the digit to its right. A terminating decimal has a finite number of digits to the right of the decimal point, while a non-terminating decimal may consist of recurring or non-recurring digits.

Denominator

The lower part of a fraction which refers to the number of equal parts into which the whole has been divided. For example, in the fraction $\frac{4}{5}$, the denominator is 5.

Diameter

The diameter is a straight line from one side of
a circle to the other that passes through the centre.

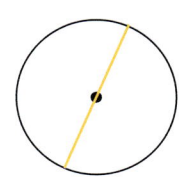

Difference

Difference usually refers to subtraction. For example, the difference between
3 and 7 is 4, since $7 - 4 = 3$.

Digit

All numbers consist of one or more digits. The digits used in the Base 10
system are 1, 2, 3, 4, 5, 6, 7, 8, 9 and 0.

Dilation

Dilations are transformations which change the size of a shape, producing
an image which is *similar* to the original object. Dilations can refer to either
enlargements or reductions.

Distributive property

When multiplying one number by another, one or both numbers may be
partitioned without affecting the result. For example, $12 \times 9 = 10 \times 9
+ 2 \times 9$.

Divisible

When one number is a multiple of another then we say that it is divisible
by this number. For example, since 24 is a multiple of 3, we say that 24 is
divisible by 3.

Dot plot

A statistical graph used to organise and display categorical data or discrete
numerical data.

Edge

The boundary line of a two-dimensional shape or the line
segment where two *faces* of a three-dimensional object meet.

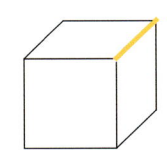

Equally likely outcomes

Equally likely outcomes have the same chance of occurring. For example, when a fair coin is tossed, heads and tails are equally likely outcomes.

Equation

A number sentence that states two expressions are equal.

Equivalent fractions

Equivalent fractions have the same relationship between the parts and the whole e.g., $\frac{1}{2} = \frac{2}{4}$.

Estimate

An approximation or judgement about the attributes of something. For example, John estimated that the length of the table was 70 cm.

Even number

An even number is an *integer* that is *divisible* by 2. The even numbers are …, −6, −4, −2, 0, 2, 4, 6, …

Expression

An expression refers to two or more numbers or variables connected by operations. Expressions do not include an equals sign.

Faces

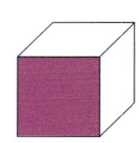

The flat surfaces of a three-dimensional object are its faces.

Factor

A factor is a whole number that divides into another number leaving no *remainder*. *Prime numbers* are numbers which have exactly two factors, while *composite numbers* have more than two factors.

Fraction

Fractions are numbers which can be written as the ratio of two integers (i.e., *rational* numbers). Fractions consist of a *numerator* and a *denominator*, separated by a *vinculum* (usually referred to as the fraction line).

Frequency

The number of occurrences of an event or the rate at which something happens.

Frequency distribution

Divides a set of observations into a number of classes according to the number of observations that fall in each class. Frequency distributions can be displayed in the form of a *frequency table*, a *two-way table* or in graphical form.

Frequency table

A frequency table lists the frequency of observations in a number of different classes, called class intervals.

Function

A function assigns to each element of a set of input values a unique element of a set of output values.

Grid reference

A grid reference identifies a region on a map. Coordinates and gridlines are used to refer to specific features or locations.

Highest common factor (HCF)

The Highest Common Factor is sometimes known as the Greatest Common Factor (GCF) or Greatest Common Divisor (GCD). The HCF is the largest common factor of two or more numbers e.g., the factors of 8 are 1, 2, 4 and 8. The factors of 12 are 1, 2, 3, 4, 6 and 12. The HCF of 8 and 12 is 4.

Histogram

A histogram displays the frequency distribution of continuous data. A histogram is a graphical representation of the information contained in a frequency table. In a histogram, class frequencies are represented by the areas of rectangles centred on each class interval. Unlike bar and column graphs, there are no spaces between adjacent rectangles.

Image

The result of applying a *transformation* (i.e., reflection, rotation or translation) to an *object*.

Improper fraction

A fraction in which the numerator is greater than or equal to the denominator.

Independent and dependent variables

When we deliberately change a variable to observe the effect it has on another variable, the variable that we change is called the independent variable. The one that we observe is called the dependent variable. For example, when we measure the temperature inside different coloured cars, the colour of the car is the independent variable and the temperature is the dependent variable.

Independent event

Two *events* are *independent* if knowing the outcome of one event tells us nothing about the outcome of the other event.

Index notation (and Indices)

When the product (such as $a \times a \times a$) is written as a^3, we say it is written in index notation, with base a. The number 3 is called the index, power or exponent.

Inequality

An inequality is a statement that one number or expression is greater than (>) or less than (<) another. For example, 5 > 3 or 11 + 2 < 15.

Integers

A number that has no decimal or fractional part (i.e., …, −2, −1, 0, 1, 2, …). Positive integers are those integers greater than zero (i.e., 1, 2, 3, …), while negative integers are those less than zero (i.e., −1, −2, −3, …). Zero is the only integer that is neither positive or negative. The set of non-negative integers consists of the positive integers and zero.

Interval

An interval is a *subset* of the *number line*. The interval from 2 and 4 is shown opposite.

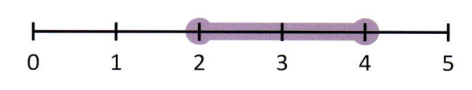

Irrational number

Fractions are numbers which can be written as the ratio of two integers (i.e., *rational* numbers). Irrational numbers are numbers which cannot be written as fractions e.g., π.

Irregular shape

Irregular shapes do not have all sides of equal length and all angles of equal size.

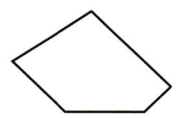

Length

Length is an attribute that refers to the distance from one end of an object to the other. The standard unit of length is the metre.

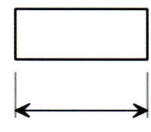

Line

A line extends infinitely in both directions. Lines are depicted with arrowheads on both ends to distinguish them from *rays* and *line segments*.

Linear equation

A linear equation is a function that produces a straight line graph.

Line graph

A diagram used to represent continuous data. The independent variable is plotted on the horizontal axis, while the dependent variable is plotted on the vertical axis.

Line of symmetry

A line about which a shape may be folded so that the two halves fit exactly on top of each other.

Line segment

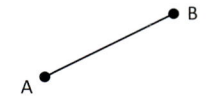

A line segment (or *interval*) extends between two points.

Lowest common multiple

The Lowest Common Multiple (LCM) of two (or more) numbers is the smallest multiple of all the numbers in the set. For example, the common multiples of 2, 3 and 4, are 12, 24, 36, … Therefore, the Lowest Common Multiple of 2, 3 and 4, is 12.

Mass

Mass is an attribute that refers to the amount of matter in an object. Mass is often confused with weight, which is the force that gravity exerts on the object. Mass is measured in kilograms.

Mean

The mean of a set of data is found by adding all of the scores and dividing by the number of scores in the set. The mean is one way of calculating an *average*.

Measures of central tendency

Measures of central tendency are statistics chosen to represent average or typical values in a set of data. The most common types of average are the *mean*, the *mode* and the *median*.

Median

The median of a data set is the middle score when the scores are placed in ascending order. The median is one way of calculating an *average*.

Mixed number

A number that contains both a whole number and a fractional part e.g., $1\frac{3}{4}$.

Mode

The mode is the most common score in a set of data. A set of data which has two modes is called *bimodal* and a set which has more than two modes is called *multimodal*. The mode is one way of calculating an *average*.

Multiples

The multiples of a number are those numbers which appear in the number's multiplication table. For example, the multiples of 6 are: 6, 12, 18, 24, 30, . . .

Natural number

A positive *integer* i.e., 1, 2, 3, … Also known as a *counting number*.

Net

A net is a two-dimensional figure that can be folded to form a polyhedron.

Non-standard units (or Informal units)

Units that are not part of a standardised system of units for measurement. For example, an informal unit for length could be paperclips of uniform length.

Number

A quantity that is made up of one or more digits. In the Base 10 system, both the value of digits and their position in the number influence the size of the number being represented.

Number line

A number line provides a visual representation of the real numbers. 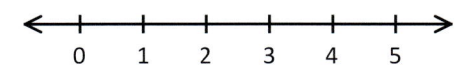 Number lines always have smaller numbers on the left and larger numbers on the right.

Number sentence

A number sentence is typically an equation or inequality expressed using numbers and symbols. For example, $2 + 3 \times 4 = 14$ or $3 \times 5 > 10$.

Numerals

Numerals are symbols used to represent numbers. For example, 4 and IV are numerals used to represent the number four.

Numerator

The upper part of a fraction which indicates how many of the equal parts are being considered.

Odd number

An odd number is an *integer* that is not *divisible* by 2. The odd numbers are …, -5, -3, −1, 1, 3, 5, …

Operation

A mathematical calculation such as addition, subtraction, multiplication or division.

Order of operations

The rules used to perform calculations. Calculations in brackets should be performed first, followed by calculations involving indices, then multiplication and division (working from left to right), and finally addition and subtraction (also from left to right).

Ordered pair

A set of two numbers whose order is significant. Ordered pairs are used to describe the location of a point in the Cartesian plane e.g., (x, y).

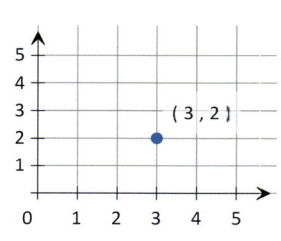

Ordinal number

A number used to indicate position, such as 1st, 2nd or 3rd.

Outlier

A data value that stands out from the other members of the set.

Parallel lines

Two lines that are always the same distance apart and which do not intersect or touch each other at any point.

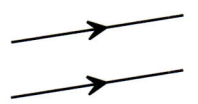

Partitioning

Dividing a quantity into parts, for example $10 = 6 + 4$.

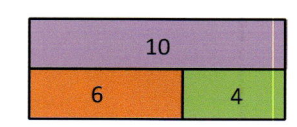

Percentage

A rate, number or amount expressed in relation to 100.

Perimeter

The distance around the outside of a two-dimensional shape. In the case of a circle, the perimeter is known as the circumference.

Perpendicular

In geometry, two lines, shapes or surfaces that meet at a right angle (90°). In this diagram, perpendicular lines are indicated by the square which represents a right angle.

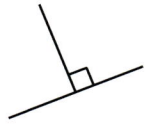

Pi

An irrational number that is the ratio of the circumference of a circle to its diameter, with a value of approximately 3.14.

Picture graph

A form of bar graph where the bars have been replaced by rows or columns of pictures of objects.

Place value

The value of a digit depends on its position in a number. For example, in 267 the 6 has a value of 60, while in 627 the 6 has a value of 600.

Point

A point marks a position in space, but has no size.

Polygon

A closed two-dimensional shape with three or more straight sides.

Polyhedron

A polyhedron is a three-dimensional object which consists of a collection of polygons joined at their edges.

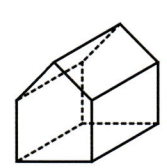

Population

A term often used in the context of statistics, population is the complete set of individuals, objects, places, etc. about which we want information.

Primary data

Primary data are original data collected by the user. Primary data might include data obtained from interviews the user has conducted personally, or observations the user makes during an experiment.

Prime factor

A factor of a number that is also a prime number e.g., while 3 and 4 are both factors of 12, 3 is a prime factor but 4 is not.

Prime number

A prime number is a number with exactly two factors: 1 and itself. For example, 5 is a prime number since 1 and 5 are its only factors. Neither 0 or 1 are prime numbers.

Prism

A solid object with two identical ends and flat sides. Prisms are named after the shape of the base. For example, a prism with a triangular base is called a triangular prism.

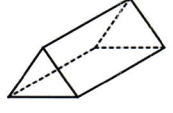

Probability

Probability refers to the chance that an event will occur. Probability is usually expressed as a fraction, decimal or percentage.

Product

A product is the result of multiplying together two or more numbers or algebraic expressions known as *factors*. Words that may indicate a product include 'of', 'per' and 'times by'.

Proof

In mathematics, proof refers to a convincing logical argument that a proposition is true.

Proportion

Proportion compares a part to the whole. Two quantities are in proportion if there is a constant *ratio* between them. For example, there are two squares for every circle.

Protractor

An instrument used to measure *angles*. It is commonly in the shape of a semicircle (180°) or circle (360°).

Pyramid

A solid object whose base is a polygon and whose sides meet at a point called the apex. Pyramids are named according to the shape of the base. This is an example of a square pyramid.

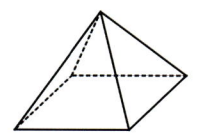

Pythagoras' theorem

Pythagoras' theorem states that for any right-angled triangle, the square of the hypotenuse (c) is equal to the sum of the squares of the lengths of the other two sides (a and b).

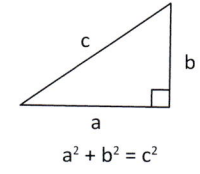

Quadrant

One of the four equal areas created when a plane is divided by two perpendicular *axes*.

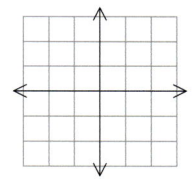

Quadrilateral

A four-sided polygon. Types of quadrilateral include squares, rectangles, parallelograms, rhombuses, trapeziums and kites.

Parallelogram	A quadrilateral with two pairs of parallel sides.
Rectangle	A quadrilateral with right angles and two pairs of parallel sides.
Rhombus	A quadrilateral with all sides equal.
Square	A quadrilateral with right angles and all sides equal.
Trapezium	A type of quadrilateral with one set of parallel sides.
Kite	Two pairs of adjacent sides that are equal in length.

Quotient
A quotient is the result of performing a division.

Radius
The radius is the distance from the centre of a circle to a point on its perimeter (known as the *circumference*). The radius is equal to half of the diameter.

Random number
A number whose value is determined by chance, such as the number of dots showing when a fair die is tossed. The value of a random number cannot be predicted in advance.

Random sample
A sample that is selected randomly from a *population*. All of the elements of the population have an equal chance of being selected in the sample.

Range
A measure of the spread of a data set. The range is the difference between the largest and smallest observations in a data set.

Ratio
A measure of the comparative relationship between two quantities. For example, the ratio of left handed people to right handed people might be 4:17.

Rational numbers
Any number which can be expressed as the ratio of two integers (i.e., a fraction).

Ray
A ray extends from a point towards infinity. Rays are usually depicted with an arrowhead which indicates the direction in which the line continues to infinity.

Real number
The numbers generally used in mathematics in scientific work and in everyday life are the *real numbers*. They can be pictured as *points* on a

number line, with the integers evenly spaced along the line. A real number 'b' is to the right of a real number 'a' if b > a. A real number is either *rational* or *irrational*. Rational numbers are those whose decimal expansions are either *terminating* or *recurring* decimals, while irrational numbers are infinite, non-recurring decimals.

Rearranging parts

Rearranging parts refers to moving counters, numbers, etc., in order to change the visual representation of the number.

Reasonableness

Reasonableness refers to how appropriate an answer is. 'Does this answer make sense?' and 'Does this answer sound right?' are two questions that should be asked when thinking about reasonableness.

Reflection

A type of transformation that involves flipping an object across a mirror line to produce the *image*.

Regular polygon

A polygon with all sides of equal length and all angles of equal size.

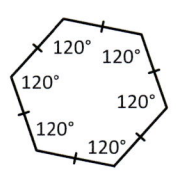

Related denominators

Denominators are related when one is a multiple of the other. Fractions with related denominators are easier to add and subtract than fractions with unrelated denominators. For example, $\frac{1}{5}$ and $\frac{3}{10}$ have related denominators.

Remainder

The amount left over when a number or quantity is divided by another. For example, 17 ÷ 5 = 3 *r* 2, since (3 × 5) + 2 = 17.

Renaming fractions

Multiplying both the numerator and the denominator of a fraction by the same non-zero number. This produces an *equivalent fraction*.

Right-angled triangle

A triangle in which two of the sides meet at a *right angle* (90°). The symbol used to denote a right angle is a square.

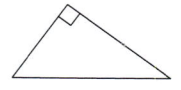

Rotation

A type of transformation that involves turning an object through a certain *angle*, either clockwise or anticlockwise, to produce the *image*.

Rounding

Reducing the number of digits in a number but keeping the value approximately the same.

Sample

In statistics, a sample refers to a subset of the *population* that is being studied.

Sample space

A sample space is the set of all possible outcomes of a chance experiment.

Scatter plot

A graphical representation of the relationship between two variables, with the horizontal and vertical axes representing the two variables. Each point plotted represents an ordered pair.

Scientific notation

A convenient way to represent very large or very small numbers. Numbers are expressed as a decimal number with just one digit to the left of the decimal point multiplied by a power of 10. For example, 6.28×10^3 represents 6 280.

Secondary data

Secondary data are data collected by others. Sources of secondary data include web-based data, the media, books, scientific papers, etc.

Sequence

An arrangement which follows a particular pattern or order.

Set

A set is a well-defined collection of objects, events or outcomes. Each item within a set is called an element of the set.

Similarity

Similar figures have the same shape but are not necessarily the same size.

Skip counting

Counting by a number that is not 1. For example, 2, 4, 6, 8, … or 10, 20, 30, …

Solid

A three-dimensional geometrical figure.

Square number

The result of multiplying a number by itself.

For example, 9 is a square number since $3 \times 3 = 9$.

Standard deviation

A measure of the variability or spread of a set of data.

Standard units

Units that are in common usage, such as metres, litres and kilograms.

Stem and leaf plot

A data display based on splitting each data value into a stem and a leaf. For example, 41 would be represented with stem 40 and leaf 1.

Subitising

Recognising how many are in a set without counting.

Subset

A subset is a set, for which all elements are part of another, possibly larger, set. For example, blue eyes are a subset of the set of eye colours.

Sum

A sum is the result of adding together two or more numbers or expressions. For example, the sum of 9 and 5 is 14. Language used to indicate a sum includes add, plus, combine, etc.

Supplementary

Two angles that add to 180° are called supplementary angles; for example, 45° and 135° are supplementary angles.

Surface area

The total *area* of all faces of a three-dimensional object.

Theorem

A mathematical statement that has been established by means of a *proof*.

Three-dimensional

An object is *three-dimensional* when it possesses the dimensions of height, width and depth. A *solid* is any geometrical object with three-dimensions.

Transformation

There are three types of transformation that change the position of shapes: *reflections* (flips), *translations* (slides) and *rotations* (turns).

Translation

A type of transformation that involves sliding an object a particular distance in a particular direction to produce the *image*.

Transversal

A line that crosses a pair of parallel lines creating angle relationships including alternate angles, various corresponding angles, co-interior angles and vertically opposite angles.

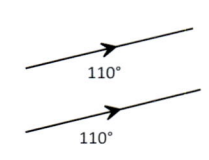

Tree diagram

A diagram which shows all of the possible outcomes of a chance experiment.

Triangular number

A number which can be expressed as a triangular array. The sequence of triangular numbers is 1, 3, 6, 10, 15, 21, … and is found by increasing the difference between successive terms by 1 each time.

Two-dimensional

Two-dimensional objects only have two dimensions: length and width.

Two-way table

A way of showing the frequency distribution when a group of individuals or things are categorised according to two criteria.

	Numbers	Letters
Green	2 5	a p
Not green	7 1 3	b c x

Unit fraction

A fraction in which the numerator is 1.
For example, $\frac{1}{3}$ and $\frac{1}{5}$ are unit fractions.

Variable

In algebra, a variable is a symbol used to represent an unspecified number of a particular type. In statistics, a variable is a measurable or observable quantity that is expected to either change over time or between observations. Examples of variables in statistics include the age of students, their hair colour or a playing field's length or its shape.

Venn diagram

A diagram that uses overlapping shapes (often circles) to show the relationships between groups that have something in common. Venn diagrams can have more than two overlapping shapes.

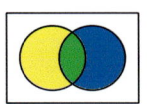

Vertex (plural Vertices)

A *point* at which two or more *edges* meet.

Vinculum

The line that separates the numerator from the denominator in a fraction.

Volume

Volume refers to the amount of space a three-dimensional object occupies. Volume is measured in cubic units, such as cubic metres or cubic centimetres. Volume is often confused with *capacity*.

Whole numbers

A number that has no fractional parts e.g., 12 is a whole number but 12.1 is not.

Answers to Exercises

Chapter 10

10.1 **Q1**

(a) Six thousand and twenty-three

(b) Twelve thousand, three hundred and fifty-seven

(c) One hundred and forty-four thousand, and fifty-seven

(d) One million, fifty thousand and sixty

Q2

(a) 423 064

(b) 8 947

(c) 1 111

(d) 19 014

Q3

(a) 6.25×10^5

(b) 4.5×10^{-4}

(c) 4.12×10^4

(d) 1.25×10^6

Q4

(a) 402 000

(b) 0.06714

(c) 98.1

(d) 3.1

Q5

(a) 14 800

(b) 2 000

(c) 0

10.2 (a) 117

(b) 590

(c) 2 056

(d) 382

(e) 284

(f) 2 913

10.3 (a) 42

(b) 55

(c) 1 848

(d) 9

(e) 365

(f) 1 119

10.4 (a) 868

(b) 1 184

(c) 21 012

10.5 (a) 1 627

(b) 293.5 or 293 r2

(c) 4 115

10.6 1. 37

3. 11

5. 39

7. 5

9. 10

2. 12

4. 19

6. 34

8. 14

10. 5

Chapter 11

11.1 (a) $\frac{1}{5}$

(b) $\frac{7}{15}$

(c) $\frac{4}{7}$

(d) $\frac{7}{13}$

(e) $\frac{3}{5}$

(f) $\frac{2}{9}$

11.2 **Q1**

(a) $\frac{2}{3}$

(b) $\frac{5}{6}$

(c) $\frac{5}{6}$

(d) $1\frac{2}{3}$

Q3

(a) $\frac{37}{45}$

(b) $\frac{19}{28}$

(c) $1\frac{1}{12}$

(d) $1\frac{7}{30}$

Q2

(a) $\frac{1}{3}$

(b) $\frac{2}{7}$

(c) $\frac{1}{2}$

(d) $\frac{3}{5}$

Q4

(a) $\frac{5}{9}$

(b) $\frac{1}{12}$

(c) $\frac{5}{24}$

(d) $\frac{7}{30}$

11.3 **Q1**

(a) $\frac{8}{15}$

(b) $\frac{5}{14}$

(c) $3\frac{3}{5}$

(d) 1

Q2

(a) $2\frac{2}{3}$

(b) 2

(c) $6\frac{1}{2}$

(d) 15

Chapter 12

12.1 (a) 12.9

(b) 216.01

(c) 0.019

(d) 49.1

(e) 0.025

(f) 999.99

12.3 (a) 4.2

(b) 0.55

(c) 1.848

(d) 0.58

(e) 3.65

(f) 1.119

12.2 (a) 5.5

(b) 5.9

(c) 20.56

(d) 38.2

(e) 8.4

(f) 4.213

12.4 (a) 0.0868

(b) 11.84

(c) 2.1012

12.5 (a) $38 \div 16$

(b) $623 \div 381$

(c) $150 \div 25$

(d) $423 \div 141$

(e) $1633 \div 880$

(f) $16 \div 80$

12.6 **Q1**

(a) 16%

(b) 42.5%

(c) 101%

(d) 0.5%

Q3

(a) $\frac{17}{50}$

(b) $\frac{7}{100}$

(c) $2\frac{2}{25}$

(d) $1\frac{19}{100}$

Q5

(a) $\frac{16}{25}$

(b) $\frac{59}{1000}$

(c) $1\frac{1}{10}$

(d) $\frac{121}{1000}$

Q2

(a) 0.38

(b) 1.418

(c) 0.67

(d) 0.0109

Q4

(a) 0.6

(b) 1.2

(c) $0.91\overline{6}$

(d) $0.4\overline{3}$

Q6

(a) 45%

(b) 62.5%

(c) 225%

(d) 42.5%

Chapter 13

13.1 (a) 70°

(b) 123°

(c) 252°

(d) 38°

13.2 **Q1**

(a) 1 250 cm

(b) 111.7 cm

(c) 3 400 m

(d) 500 cm

13.2 **Q2**
    (a) 12 050 g
    (b) 1.355 t
    (c) 0.468 kg
    (d) 25 g

Q3
    (a) 0.12 kL
    (b) 3.45 L
    (c) 60 L
    (d) 45 500 mL

13.2 **Q4**
    (a) 56 days
    (b) 72 hours
    (c) 450 minutes
    (d) 8 760 hours

Q5
    (a) 0.0001 km^2
    (b) 10 cm^2
    (c) 2.5 m^2
    (d) 5 000 m^2

Chapter 14

14.1 (a) 24 cm
      (b) 24 cm
      (c) 23.42 cm
      (d) 24 cm

14.2 (a) 24 cm^2
      (b) 24 cm^2
      (c) 38.13 cm^2
      (d) 35 cm^2

14.3 (a) 90 cm^3
      (b) 48 cm^3
      (c) 40 cm^3
      (d) 64 cm^3

14.4 (a) 785 cm^3
      (b) 339.12 cm^3
      (c) 1 570 cm^3
      (d) 452.16 cm^3

14.5 (a) 126 cm^2
      (b) 108 cm^2

Chapter 15

15.1 (a) $88 = 2^3 \times 11$
      (b) $120 = 2^3 \times 3 \times 5$
      (c) $400 = 2^4 \times 5^2$
      (d) $960 = 2^6 \times 3 \times 5$
      (e) $1\ 080 = 2^3 \times 3^3 \times 5$
      (f) $1\ 296 = 2^4 \times 3^4$

15.2 **Q1** 16, 22, 29

      Q2 $-\frac{1}{2}, -\frac{1}{4}, -\frac{1}{8}$

      Q3 $7\frac{1}{2}, 8\frac{3}{4}, 10$

      Q4 0.0005, 0.00006, 0.000007

      Q5 3.125, 1.5625, 0.78125

      Q6 364, 1 093, 3 280

      Q7 35, 48, 63

      Q8 16, 26, 42

      Q9 161 051, 1 771 561, 19 487 171

      Q10 111 111, 1 111 111, 11 111 111

      Q11 85, 79, 72

      Q12 $1\frac{1}{8}, 1\frac{1}{4}, 1\frac{3}{8}$

      Q13 $1\frac{31}{32}, 1\frac{63}{64}, 1\frac{127}{128}$

      Q14 $\frac{6}{7}, \frac{7}{8}, \frac{8}{9}$

      Q15 11.0, 12.1, 13.2

Chapter 16

16.1 **Q1**

    (a)   171.4 cm

    (b)   317 g

    (c) $8.98\bar{3}$

    (d)   \$2 650 000

    Q3

    (a)   9

    (b)   6

    (c)   4.5

    (d)   0.25

Q2

(a) 1

(b) The modes are 0.4 and 0.5

(c) 14

(d) 2

Chapter 17

17.1 (a) $P(black) = \frac{5}{20}$ 17.2 (a) $P(blue) = \frac{1}{8}$

  (b) $P(red\ or\ green) = \frac{5}{20}$ (b) $P(red\ or\ green) = \frac{5}{8}$

  (c) $P(yellow) = \frac{0}{20}$ (c) $P(yellow) = \frac{2}{8}$

  (d) $P(not\ green) = \frac{18}{20}$ (d) $P(not\ green) = \frac{6}{8}$

17.3 (a) {0, 1, 2, 3, 4, 5, 6, 7, 8, 9}

  (b) {A, E, I, O, U}

  (c) {Jan, Feb, Mar, Apr, May, Jun, Jul, Aug, Sep, Oct, Nov, Dec}

  (d) {M, Tu, W, Th, F, Sa, Su}

  (e) {1, 2, 3, 4, 6, 12}

  (f) {WA, SA, QLD, NSW, VIC, TAS}

17.4

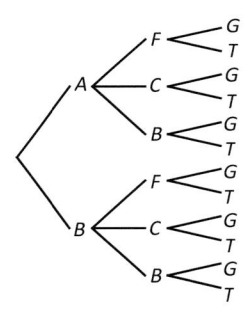

There are 12 possible meal options.

17.5 (a), (b), (d) and (e) are independent events.

17.6 (b), (c), (e) and (f) are mutually exclusive.

References

Ainley, J., Kos, J. & Nicholas, M. (2008). Participation in science, mathematics and technology in Australian education. Retrieved from http://research.acer.edu.au/acer_monographs/4.

Artemenko, C., Daroczy, G. & Nuerk, H.-C. (2015). Neural correlates of math anxiety—an overview and implications. *Frontiers in Psychology, 6.* doi:10.3389/fpsyg.2015.01333

Australian Council for Educational Research [ACER]. (2017). Literacy and numeracy test for initial teacher education students: Assessment Framework. Retrieved from https://teacheredtest.acer.edu.au/.

Australian Curriculum, Assessment and Reporting Authority [ACARA]. (2017). The Australian Curriculum: Mathematics. Retrieved from www.australiancurriculum.edu.au/.

Australian Institute for Teaching and School Leadership [AITSL]. (2014). Australian Professional Standards for Teachers. Retrieved from www.aitsl.edu.au/teach/standards.

Bandura, A. (1989) Social cognitive theory. In R. Vasta (Ed.), *Annals of child development Vol. 6. Six theories of child development* (pp. 1–60). Greenwich, CT: JAIPress.

Baroody, A.J. & Coslick, R.T. (1998). *Fostering children's mathematical power: An investigative approach to mathematics instruction*. Mahwah, NJ: Lawrence Erlbaum Associates.

Barry, K. & King, L. (1998). *Beginning teaching and beyond* (3rd edn). Melbourne, VIC: Social Science Press.

Boaler, J. (2009). *The elephant in the classroom: Helping children learn and love maths*. London: Souvenir Press.

Boaler, J. (2014). Fluency without fear: Research evidence on the best ways to learn math facts. Retrieved from www.youcubed.org/wp-content/uploads/2015/03/FluencyWithoutFear-2015.pdf.

Boaler, J. (2015). *What's math got to do with it?* (2nd edn). New York, NY: Penguin.

Boaler, J. (2016). *Mathematical mindsets: Unleashing students' potential through creative math, inspiring messages and innovative teaching*. San Francisco: Jossey-Bass.

Bobis, J., Mulligan, J. & Lowrie, T. (2013). *Mathematics for children: Challenging children to think mathematically* (4th edn). Frenchs Forest, NSW: Pearson.

Booker, G., Bond, D., Sparrow, L. & Swan, P. (2004). *Teaching primary mathematics* (3rd edn). Frenchs Forest, NSW: Pearson.

Brady, K. & Winn, T. (2017). *Maths skills for success at university*. South Melbourne, VIC: Oxford University Press.

Brown, G. (2009). *Review of education in mathematics, data science and quantitative disciplines*. Turner, ACT: The Group of Eight.

Charles, C.M. (1999). *Building classroom discipline* (6th edn). New York, NY: Longman.

Charles, C.M. (2011). *Building classroom discipline* (10th edn). Boston, MA: Pearson.

Dweck, C.S. (2017). What is mindset. Retrieved from https://mindsetonline.com/whatisit/about/index.html

Hackling, M., Murcia, K., West, J. & Anderson, K. (2014). *Optimisation of STEM Education Support in Western Australian Schools*. Perth, Western Australia: Department of Commerce.

Harris, K. & Jensz, F. (2006). *The preparation of maths teachers in Australia*. Melbourne, VIC: Australian Council of Deans of Science.

Hattie, J. (2012). *Visible learning for teachers*. New York: Routledge.

Hattie, J. (2016). *Visible learning for mathematics, grades K–12: What works best to optimize student learning*. Thousand Oaks, CA: SAGE Publications.

Haylock, D. (2006). *Mathematics explained for primary teachers* (3rd edn). London, UK: SAGE Publications.

Haylock, D. & Manning, R. (2014). *Mathematics explained for primary teachers* (5th edn). London: SAGE Publications.

International Association for the Evaluation of Educational Achievement (IEA) (2016). TIMSS 2015 international results in mathematics. Retrieved from https://nces.ed.gov/timss/educators.asp.

Khan, S. (2012). *The one world school house: Education reimagined*. New York: Hachette Books Group.

Khan Academy. (2017). About Khan Academy. Retrieved from www.khanacademy.org/about.

Krause, K., Bochner, S. & Duchesne, S. (2003). *Educational psychology for learning and teaching*. Southbank, VIC: Nelson.

McConney, A. & Price, A. (2009). Teaching out-of-field in Western Australia. *Australian Journal of Teacher Education, 34*(6). doi:10.14221/ajte2009v36n6.6.

Mullis, I., Martin, M., Foy, P. & Arora, A. (2012). *TIMSS 2011 international results in mathematics*. Amsterdam, The Netherlands: IEA.

Office of the Chief Scientist (2012). *Mathematics, engineering and science in the national interest*. Retrieved from www.chiefscientist.gov.au/wp-content/uploads/Office-of-the-Chief-Scientist-MES-Report-8-May-2012.pdf.

Ojose, B. (2008). Applying Piaget's theory of cognitive development to mathematics instruction. *The Mathematics Educator, 18*(1), 26–30.

Reys, R., Lindquist, M., Lambdin, D., Smith, N., Rogers, A. et al. (2012). *Helping children learn mathematics: 1st Australian Edition*. Milton, QLD: John Wiley & Sons Australia.

Shulman, L.S. (1986). Those who understand: Knowledge growth in teaching. *Educational Researcher, 15*(2), 4–14.

Siemon, D., Beswick, K., Brady, K., Clark, J., Faragher, R. & Warren, E. (2011). *Teaching mathematics: Foundations to middle years*. South Melbourne, VIC: Oxford University Press.

Sparrow, L. & Swan, P. (2005). *Starting out: Primary mathematics*. Prahran, VIC: Eleanor Curtain Publishing.

Sullivan, P. (2011). *Teaching mathematics using research-informed strategies*. Camberwell, VIC: ACER.

Swan, P. & Marshall, L. (2010). Revisiting mathematics manipulative materials. *Australian Primary Mathematics Classroom, 15*(2), 13–19.

Teacher Registration Board of Western Australia [TRBWA]. (2016). Literacy and numeracy test for initial teacher education students. Retrieved from http://www.trb.wa.gov.au/Pages/Literacy-and-Numeracy-Test-for-Initial-Teacher-Education-Students.aspx

Thomson, S., De Bortoli, L. & Buckley, L. (2013). PISA 2012: How Australia measures up. Retrieved from www.acer.edu.au/documents/PISA-2012-Report.pdf.

Thomson, S., Wernert, N., Underwood, C. & Nicholas, M. (2008). *TIMSS 2007: Taking a closer look at mathematics and science in Australia*. Retrieved from http://research.acer.edu.au/timss_2007/2/.

Westwood, P. (2000). *Numeracy and learning difficulties: Approaches to learning and assessment*. Melbourne, VIC: ACER.

Image Acknowledgements

p. 22   pan-xiaozhen, Unsplash

p. 23   dreamerve, Shutterstock

p. 24   ImageFlow, Shutterstock

p. 26   Pyty, Shutterstock

p. 28   Pressmaster, Shutterstock

p. 31   StartupStockPhotos, Pixabay (top); rawpixel, Unsplash (bottom)

p. 32   rawpixel, Unsplash (top); Thought Catalog, Unsplash (bottom)

p. 35   goldyg, Shutterstock

p. 45   Morrowind, Shutterstock (top); k_samurkas, Shutterstock (bottom)

p. 48   Poznyakov, Shutterstock

p. 50   Monkey Business Images, Shutterstock

p. 62   spass, Shutterstock

p. 63   (top to bottom) John West, personal collection; Ignasi Soler,
          Shutterstock; Antonia Giroux, Shutterstock; Jesse Davis,
          Shutterstock; CJM Grafx, Shutterstock; John West, personal
          collection; Iryna Imago, iStock; weerastudio, Shutterstock

p. 77   Mimi Thian, Unsplash

p. 80   TJ Evans, Pixabay

p. 82   Tom Wang, Shutterstock

p. 90   Myriams-Fotos, Pixabay

p. 92   Thought Catalog, Unsplash (top); rawpixel, Unsplash (bottom)

p. 93   kreatikar, Pixabay (top); rawpixel, Unsplash (bottom)

p. 94   Kelly Sikkema, Unsplash

p. 104   Mariamichelle, Pixabay

p. 135   Jarmoluk, Pixabay

p. 136   meunierd, Shutterstock

p. 152   Aghilms, Pixabay

p. 155   Victor Freitas, Unsplash (top); Veri Ivanova, Unsplash (bottom)

p. 156   OpenClipart-Vectors, Pixabay

p. 173   Ruston Youngblood, Unsplash

p. 174   Charles Deluvio, Unsplash

p. 201   mcmurryjulie, Pixabay

Index